CATTLE FOOTCARE and CLAW TRIMMING

The origin and prevention of the necrotising inflammations of the corium (ulcerations of the claw)

E. TOUSSAINT RAVEN, DVM

Department of Large-Animal Surgery
Faculty of Veterinary Medicine
University of Utrecht
The Netherlands

Including chapters by:
R. T. HAALSTRA, DVM, PhD (Ch. 5, 'Nutrition')
D. J. PETERSE, DVM, PhD (Ch. 7, 'Breeding')
Faculty of Veterinary Medicine
University of Utrecht

Illustrated by:
ALOYS LURVINK
Educational Media Institute
University of Utrecht

Foreword by:
P. J. N. PINSENT, BVSc, FRCVS
Department of Veterinary Medicine
University of Bristol

FARMING PRESS BOOKS

Acknowledgements

The publisher and author acknowledge with thanks the assistance given by Mr P.J.N. Pinsent in the preparation of the English-language edition.

Photographs, pages 28 and 29, from Thesis Zantinga, 1968.

Diagrams, page 102, van Amerongen, 'Directions for the Use of Rubber Claw Blocks'.

Diagrams, page 105, Practical Centre for Dairy Cattle and Grassland Management, Oenkerk, The Netherlands.

English edition first published 1985
Third impression (with amendments), 1989
Reprinted 1992, 1995, 1997

Translated by Mrs F.M. Enserink
from the Dutch *Klauwverzorging Bij Het Rund*
originally published by Rijksuniversiteit te
Utrecht, 1977 and by Uitgeverij Terra,
Zutphen, 1981

ISBN 0 85236 149 1

Published by Farming Press
Miller Freeman Professional Ltd
Wharfedale Road
Ipswich IP1 4LG, United Kingdom

Distributed in North America
by Diamond Farm Enterprises
Box 537, Alexandria Bay
NY 13607, USA

Typeset by Galleon Photosetting, Ipswich

Printed and bound in Great Britain by
Biddles Ltd, Guildford and King's Lynn

FOREWORD

by P. J. N. PINSENT, BVSc, FRCVS
Senior Lecturer in Veterinary Medicine,
University of Bristol

IT HAS given me great pleasure to write the foreword to Mr Toussaint Raven's book. During the last twenty-five years, in association with the development of loose housing systems and with intensive feeding for high milk yields, foot lameness in the dairy cow has become increasingly important. In fact, in the United Kingdom it has now joined infertility and mastitis to become the third major disease of the dairy industry, and is responsible for a loss estimated at £35,000,000 annually. The etiology of foot lameness is varied and includes faults of conformation, nutritional excess, and environmental stress due to unsuitable concrete, the erosive effects of slurry, and the high bacterial population present in many yard and cubicle units.

In recent years a great deal of work has been done on lameness and much is now known of the nutritional and managemental factors predisposing to foot lameness; but although it has long been customary to emphasise the value of competent foot trimming in the control of the disease, there has been no authoritative work on the theory and practice of preventive footcare as a science in itself. Foot trimming, particularly of the hind feet, is an all-important technique, for undoubtedly the health and proper functioning of the dairy cow's foot depends to a very large extent upon the maintenance of its natural and normal shape, so that the correct weight distribution is always present.

Mr Toussaint Raven's book is all about trimming, and covers every aspect of the care of the cow's foot.

There are sections on the basic anatomy and physiology of the structures of the foot, leading to a consideration of the bio-mechanics of locomotion and their effect upon the stresses to which the feet are subjected in the environment of the modern dairy cow.

The various disease conditions causing foot lameness are discussed in detail, particularly from the etiological standpoint. No attempt is made to discuss medical, nor the more sophisticated surgical, treatments available for the cure of these diseases. The whole theme of the book is that the proper balance and weight distribution obtained by competent and regular trimming can, by itself, both prevent and cure most foot disease. The sections on the technique of foot trimming are fascinating and must be of great interest to cattle veterinarians and progressive farmers alike.

The diagrams and illustrations which appear in profusion throughout the book are superb. They are, in fact, almost self-explanatory, and there is no doubt in my mind that the book would be worth having for the diagrams alone, even were no text supplied.

There are sections describing the tools of the trade and excellent advice is available on the correct methods of sharpening foot knives. Even after many years of experience in foot trimming the reader will find a considerable improvement in results if he or she takes heed of Mr Toussaint Raven's advice on knife sharpening techniques.

All in all this is a fascinating and informative book, which will be as useful to the experienced foot-trimmer as to the student beginning to learn. It should, however, be remembered that Mr Toussaint Raven has written his book in the environment of the Dutch dairy industry, and has left the reader to make the necessary amendments to bring the information it contains into line with prevailing British conditions, where different housing systems and management practices can make a considerable difference to the stresses affecting the feet. For instance, the foot disease interdigital dermatitis (heelhorn erosion) to which the author pays a great deal of attention and devotes much space, is very conditional on the housing system and the way this functions and is practically unknown in this country, at least in its typical form. It is possible that the condition known here as *Erosio ungulae* (slurry heel) may be a mild chronic form of the Dutch disease, but the condition is only potentially of importance here.

The same applies to the condition of digital dermatitis, which is practically unknown here at the present time.

Nomenclature can occasionally be confusing. It would be quite natural to believe that the condition of interdigital dermatitis is our 'foul-in-the-foot' or 'loor'. It is not. Mr Toussaint Raven uses the term interdigital phlegmon to describe 'foul-in-the-foot', and although he is, of course, conforming correctly to international convention when so doing, the terminology may cause confusion in the mind of a British reader unless he concentrates carefully.

The most important foot lesions in British dairy cattle are the white line lesion and the solar ulcer. It is a pity, in our view, that the author dismisses the white line lesion as a minor manifestation of laminitis, for we believe that friction (i.e. concrete) is also very important in this condition.

Mr Toussaint Raven stresses a connection between heelhorn erosion (interdigital dermatitis) and solar ulcer. Such a connection is relatively insignificant in Britain, where solar ulcers are very commonly laminitic in origin, although we accept and support the contention that abnormal weight distribution producing a 'typical sole defect' is a vital part of the etiological pattern. We also believe that a relative ischaemia in the hind feet, due to the restriction of movement in uncomfortably short and narrow cubicles, may also play a part.

Many readers will feel that a fuller discussion of the etiological factors concerned in the development of laminitis would have been helpful, for laminitis is an extremely important basic reason for a number of diseases of the foot. Nevertheless one should remember that the book is about foot trimming and care: not about foot diseases. It is interesting, in any case, to speculate regarding the future of laminitis as the quota has its effect upon nutritional programmes.

Another interesting difference between Dutch and British practice is the use of the 'stand-in' foot bath in Holland as described in the text. Such foot baths are of course rare in this country where conditions favour the 'walk-through' type.

In spite of differences in disease problems and nomenclature between Holland and Britain Mr Toussaint Raven's book remains a very valuable addition to the literature on the bovine foot, and will inevitably do a great deal to improve the present generally low standard of foot trimming and care in this country.

September 1984   P. J. N. PINSENT

North American Usage

North American readers should note that *stable footrot* is synonymous with heelhorn erosion (interdigital dermatitis), and *pasture footrot* with foul-in-the-foot (interdigital phlegmon).

AUTHOR'S PREFACE

This manual for the care and treatment of cattle feet deals with the Dutch way of handling a problem that occurs not only in the Netherlands, but in a great many other countries as well – the *problem of claw lameness* in intensive dairy farming. Although opinions differ about several aspects of the subject, especially internationally, sufficient practical data and theoretical considerations have been gathered to tackle the question. Cattle must be kept on their feet despite foot problems; something must be *done*. Hence the following book. It is meant to provide answers to urgent practical questions in dairy farming.

Cattle Footcare has been written for veterinary students, veterinary surgeons and claw-trimmers, whether practising or training, and for farmers who are interested in the care and treatment of feet. This does not mean that everybody should read and digest the entire book. Farmers, trimmers and veterinarians will each find different points of emphasis and, in particular, certain aspects of disease will be of interest to different readers.

This publication is a survey of current knowledge and experience and it also includes some theoretical analysis. No sharp distinction will be made between the elements of knowledge, experience and analysis. The result is not an exhaustive scientific treatise, but rather a practical guide to what may happen, and what at any given moment can be done about it; what reactions can be made to a developing situation in the relationship between the foot and its environment.

This book does not deal with *all* the facts known about feet. The material is mainly practical: only those factors contributing to the care and treatment of feet will be discussed.

Since this publication is intended for practical use, and not as a scientific work of reference, the text will not refer to other articles or books with identical or different opinions. A sample of the most relevant literature can be found at the end of the book.

Cattle Footcare was written at the Faculty of Veterinary Medicine, University of Utrecht, and is based on independent research and experience. The most important findings and opinions of others have been taken into consideration.

The book is now being used as teaching material at the faculty itself and in agricultural education. It is based on work carried out on Dutch Friesian cattle.

The diagrams, and the advice to draw your own, are meant to stimulate the reader's imagination. They are also intended to elucidate the text; they should be 'read' together with the text.

In passing on knowledge and technique, human contact still has irreplaceable value. This applies to students, for whom this book is *merely an aid*. They must have opportunity to ask questions based on practical examples. It also holds true for the teacher, who must at all times be open to difficult questions and constructive criticism.

Warning

The problem of foot lameness is due to several factors, so it is not certain whether difficulties similar to the Dutch ones arise in other climates which have different housing conditions and feeding habits, and where other breeds may be involved. This must be investigated locally. Solutions will presumably not be the same everywhere; they must be adjusted to local circumstances.

Trimming can play a major part in the treatment of individual cases of lameness. If there is a herd problem, trimming can also play an important part in preventing further outbreaks.

If there is no lameness problem trimming can produce it!

Further Training

It cannot be too strongly emphasised that proficient training skills can only be learned by supervised practice and training. An illustrated text such as *Cattle Footcare and Claw Trimming* is not sufficient on its own – however useful it may be as an adjunct to correct training.

It is warmly recommended that bodies involved in establishing training schemes should contact the Practical Training Centre for Dairy Cattle and Grassland Management at Oenkerk, Netherlands.

CONTENTS

Page

Foreword by P. J. N. Pinsent   3
Preface   5
Introduction   **8**

1. Structure and Functions of the Claw   **13**
Functions of the claw.   13
The horny shoe and its bearing function   14
stability – side view – rear view – front view –
bottom view.
Protective function and thickness of the sole.   19
Inner structure and bearing function   20
biomechanics – digital cushion; corium; germinal
layer – loading of the quick – man-made conditions.
Differences between the claws   26
biomechanics – functional anatomy – enlarged outer
hind claw – typical sole-lesion – measuring weight-
bearing.
Posture of the hind legs.   32
Fore claws.   34

2. Diseases of the Quick   **35**
Disease as a herd problem.   35
Structure of the skin   36
horn formation.
Interdigital dermatitis (heelhorn erosion)   38
bacteroides nodosus – two stages – stage 1
(inflammation of the interdigital skin; defective
horn formation in the bulb area; excessive horn
formation – stage II – events in the second stage –
vicious circle – ulceration of the sole – deviating
postures and lameness – simple ulceration of the
sole – complicated ulceration of the sole –
interdigital overgrowth – fore claws – incidence
and control.
Laminitis   53
chronic laminitis – second stage – stage I of
laminitis – multifactorial disease – fore claws –
acute laminitis – subacute laminitis – fracture of the
horny sole – treatment – incidence and prevention.

Other disorders of the foot   66
interdigital phlegmon (foul-in-the-foot) – stones
congenital deformities of the claws – superficial
purulent inflammation of the corium – sandcracks –
digital dermatitis – unspecific inflammations of the
skin.
The connection between claw disorders.   74

3. Trimming   **75**
Aims and potential   75
functional trimming.
Working method   79
fixing the leg – manual lifting without aids – how to
handle pincers and hoof knife.
Examination   82
length – thickness of the sole – shape – height.
Work plan for hind claws   87
cutting the inner claw – paring the inner claw –
cutting and paring the outer claw – slope of the sole.
Lameness.   92

Flat claws.   93
Overgrowth of the interdigital skin.   93
Posture of the legs.   93
Fore claws.   94
Treatment of horn lesions.   95
Care after trimming   98
simple open lesions – protection – blocks –
instructions after trimming – treatment of
complicated conditions.
Preventive trimming.   103
Summary of treatment   105
functional trimming – horn lesions – planning.
Grinding the hoof knife   106
requisites – grinding.

4. Foot Baths   **107**
Notes on the control of heelhorn erosion.   107
Formalin.   109
Using foot baths   109
stand-in footbaths – walk-through footbaths.

5. Nutrition   **114**
Disease and laminitis.   114
Prevention of laminitis.   115

6. Housing   **117**
Housing conditions and: 117
disease – lameness – trimming – hygiene –
fore claws.
Adjustment and the critical period.   120

7. Breeding   **121**
Quality of the claw – heredity.   121
Breeding and claws.   121
Characteristics for claw quality.   122
Heritability of the causes of lameness.   122
Differences between breeds.   123
Conclusion.   123

Epilogue   **124**
The multifactorial nature of the problem   124
choices.
Planning for change.   124

Further Reading   **125**

Index   **126**

Using Rubber Claw-Blocks   **127**

INTRODUCTION

Footcare encompasses all the means and methods in dairy farming that help produce the optimal functioning of the **horny shoe** of the cow. The shoe protects the end of the leg, enabling the animal to carry its own body-weight and to move from one place to another. The shoe must be *intact* and have a *good bearing* (or supporting) *surface*, so that the cow can both stand and walk.

The same applies to shoes worn by human beings: these should be intact and have a good supporting surface; they should not be worn out or have a heel missing.

Although a human shoe and a bovine claw both give protection and enable men and cows to stand and walk about freely, the similarity between them ends here. The human shoe does not have any further relationship with the foot it contains – it exists independently of the living tissue it protects, and only breaks down after the wear and tear of daily use. A worn-out shoe can be worn on a sound foot and a good shoe can be worn on a weak or diseased foot.

On the other hand the claw, the *horny shoe*, does have a further intimate relationship with the living tissue (**the quick**), because the shoe itself is *produced* by this living tissue, just as human nails are formed by the living tissue they protect.

A healthy quick will form a healthy, strong, intact horny shoe. Moreover the formation-process of the shoe will *adapt* to daily usage, to normal wear and tear. The horny shoe will remain efficient, enabling the cow to stand and move about. If this were not the case, the cow would have been extinct long ago.

It can also be assumed that an unhealthy quick will produce a poor horny shoe; possibly deformed or not intact, but certainly inefficient. Furthermore it may be questioned whether the horn formation will be able to continue to adapt to daily usage.

The following must be kept in mind: the claw is a direct product of the living tissue it contains and which is protected by it. The quick is not protected passively; it has to create its own protection.

A well-shaped, strong claw indicates a healthy quick. This claw has a reasonable resistance to outside influences and can, within reasonable limits, adapt to changing conditions.

If the horny shoe is misshapen and defective, an unhealthy, malfunctioning quick will be suspected. What else could produce these defects, given acceptable conditions?

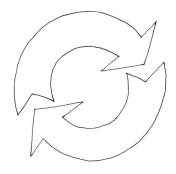

If the health of the quick is of primary importance in the formation of a protective shoe, it is clear that an unhealthy quick can no longer efficiently protect itself. As damage further worsens the already poor condition of the quick, a vicious circle is set up.

The two-way relationship between the living tissue and the horny shoe will be the starting-point for understanding bovine feet and practising footcare.

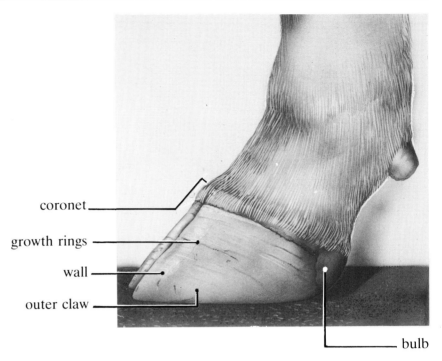

coronet

growth rings

wall

outer claw

bulb

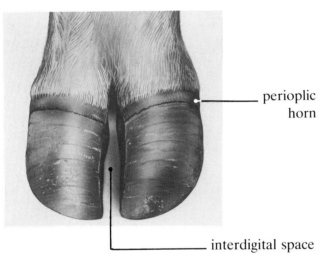

perioplic horn

interdigital space

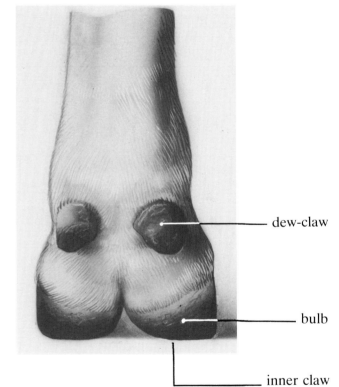

dew-claw

bulb

inner claw

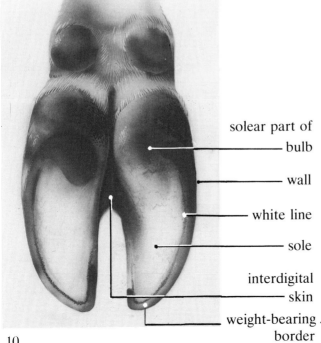

solear part of bulb

wall

white line

sole

interdigital skin

weight-bearing border

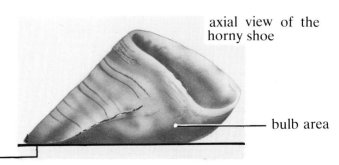

axial view of the horny shoe

bulb area

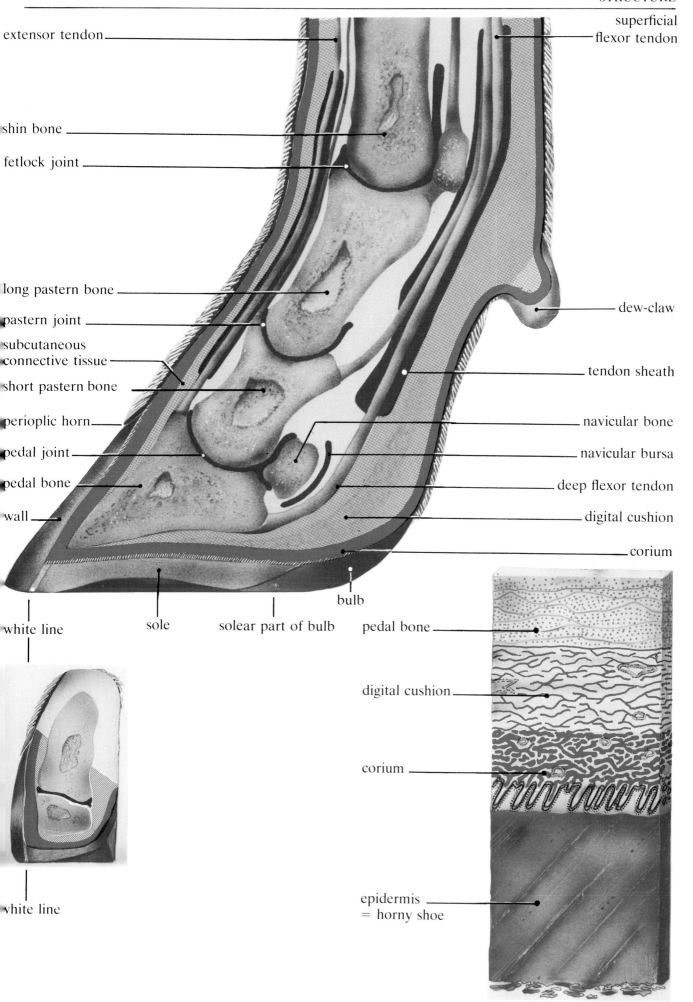

extensor tendon

superficial
flexor tendon

shin bone

fetlock joint

long pastern bone

dew-claw

pastern joint

subcutaneous
connective tissue

short pastern bone

tendon sheath

perioplic horn

navicular bone

pedal joint

navicular bursa

pedal bone

deep flexor tendon

wall

digital cushion

corium

white line

bulb

sole

solear part of bulb

pedal bone

white line

digital cushion

corium

epidermis
= horny shoe

11

Chapter 1

STRUCTURE AND FUNCTIONS

FUNCTIONS OF THE CLAW

The function of the claw is twofold:

- to protect the 'quick' . . . while
- carrying the body-weight.

The protection of the quick requires a strong and intact horny shoe.

The body-weight has to be carried in both rest and motion. The claw therefore is built in such a way that it can receive and bear this weight well.

A vital key to understanding something about claws and their function is what they look like and how they are constructed.

Actually doing one's own drawing is a great help here, particularly if it includes both the external shape and the internal cross-section, so that shape and contents are brought together. The relationship between structure and function will be appreciated and a functional image will be formed; almost a functioning model.

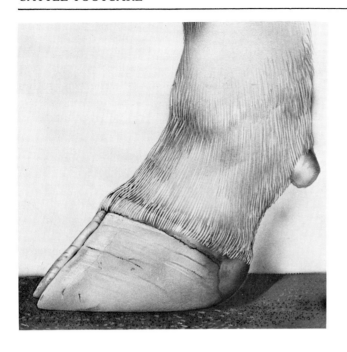

THE HORNY SHOE AND ITS BEARING FUNCTION

The diagram shows the **wall** of the **outer claw** and part of its **heel** or **bulb**, along with the anterior margin of the **inner claw**. Between the claws is the **interdigital space**, which lies along the axis of the foot, in a so-called **axial** position.

The wall can be compared to the human nail. It is continuously formed at the top (the **coronet**) and moves downwards to the bottom, where it wears off on its **weight-bearing border**. The degree of wear depends on the conditions (e.g. housing). The wall is the hardest and toughest part of the claw.

The coronet is also the area of transition between the hairy skin and the horny shoe. The transition is formed by the soft horn of the **periople**.

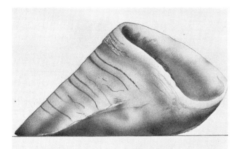

This horn partly covers the wall and can be compared to the human cuticle. It often moves slightly downwards with the horn of the wall like a flaky, dried-up layer of horn.

At the back the periople develops into the heel; the horn of the periople becomes the horn of the heel.

The visible part of the wall is the so-called **abaxial** area which is on the outside of the claw away from the interdigital space. The axial part of the wall, bordering on the interdigital space is hardly visible at all. The irregular posterior margin of the axial wall runs upwards and backwards; that is why the weight-bearing border in the interdigital space is located only in the anterior part of the claw.

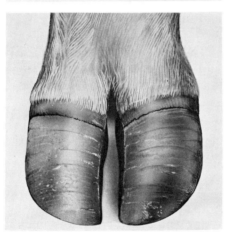

Growth rings are fine, regular stripes in the horn of the wall which run roughly parallel to the coronet. They are related to the varying speed of horn formation. The growth rings are less regular in the axial region of the wall. This area often has a very uneven surface in contrast to the smooth surface of the abaxial part of the wall.

The horn of the wall encloses the anterior part of the claw on the vertical sides. Here the bottom is formed by the **horn of the sole**, which is softer and less tough than the horn of the wall. The sole of the claw has a slightly hollow shape, its slope getting increasingly steep towards the interdigital space.

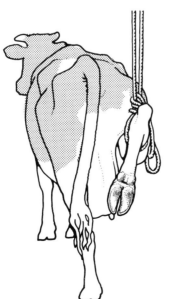

The posterior part of the claw is totally enclosed by the **horn of the heel** (or **bulb**). This includes the area at the bottom of the claw which is level with the weight-bearing border, and which could be referred to as the solear part of the heel.

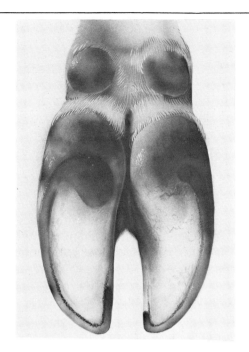

Like the perioplic horn the horn of the heel partly covers the harder horn of the wall and that of the sole. The transition is often visible as a thickened rim. Only at the slope of the sole, the horn of the sole and that of the heel merge together gradually and imperceptibly. The horn of the heel is softer and more elastic than that of the sole.

The **interdigital skin** connects the claws in the interdigital space. In the cow this skin is hairless.

The connection between the weight-bearing border of the wall and the horn of the sole is formed by the so-called **white line**. This line of horn, running along the inside of the weight-bearing border is not white but greyish. In horses this rim of softer horn is where the nails are driven in when they are shod.

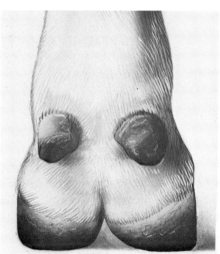

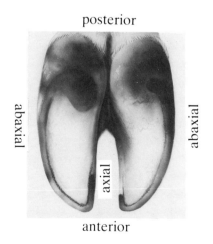

posterior

abaxial

axial

abaxial

anterior

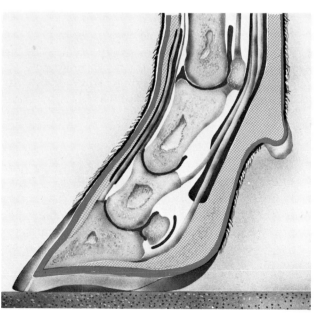

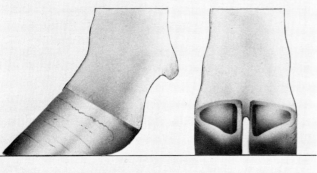

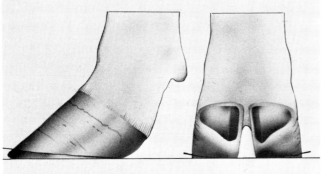

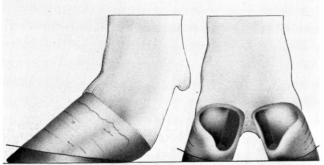

Stability

On a smooth surface the claw rests on the weight-bearing border of the wall (largely abaxially and to a small extent axially) and on part of the horn of the heel. The border of the wall and the solear part of the heel form the so-called **weight-bearing surface** or **supporting surface** of the claw. Although this is the case on a smooth hard floor, under more natural conditions, such as grass, the sole will also be part of the weight-bearing surface.

Claw growth is mainly in the anterior abaxial wall; there is less growth in the heel and axial wall. This is due to the shape of the claw and the direction in which horn formation takes place. Thus the claw becomes *longer* in the toe area, especially abaxially, but at the same time *only slightly higher* in the heel.

If this is the case when the claw is loaded on a hard surface, it will lean over axially and backwards, especially when there is little or poor development of the solear part of the heel on the axial side. The supporting surface in this case is *not stable*: the claw tilts over under the strain.

If the wall is short, and the solear part of the heel is well developed, the claw can stand firm and upright on the ground and thus the supporting surface is then *stable*.

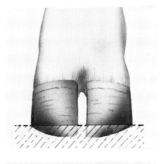

Nevertheless, under natural conditions such as uneven, springy ground, an unstable weight-bearing surface can give stable support: in an upright position, the claw can bear the body-weight, with the long wall sticking into the ground.

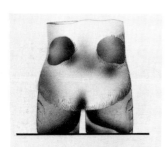

There will also be stability when claws are evenly worn off on hard ground.

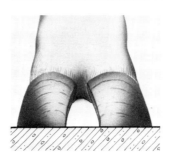

Instability will really become noticeable when the cow is confined for a long time on a hard, flat surface (e.g. when housed), *if the wall does not wear off sufficiently*.

Stability is important because the claw bears the body-weight most efficiently when it is upright beneath the body. The structure will make this clear: functionally, a claw ought to stand upright!

The trimmer's first requirement is a knowledge of the *exact shape* of a claw. To achieve this he must appreciate the outline from precise positions.

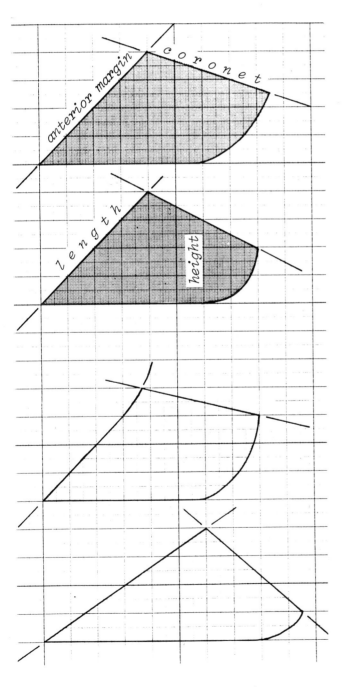

Side View

The essential elements of the side view are the slope of the (more or less straight) anterior margin, the slope of the coronet and the height of the heel. There is normally a slight difference between fore and hind claws, with the fore claw often looking somewhat steeper. However, the difference seems to be due to the heel being slightly higher rather than to any difference in the slope of the anterior margin; the coronet is more horizontal. This should not be expressed in exact measurements; it is a matter of impression, of visualisation. Trimming cannot be measured – one has to have an eye for it.

If this is a hind claw, the heel is higher than normal; the claw is too high and the coronet too level.

This hind claw is too long and too low; too long in the toe and too low in the heel. It is far too pointed and the coronet is too steep.

17

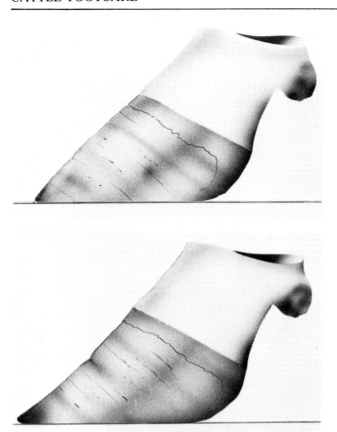

In heifers before the first calving the anterior margin of the claws usually has a slightly convex profile. After calving this profile becomes somewhat concave, especially in the outer hind claw (see under *laminitis*, page 59).

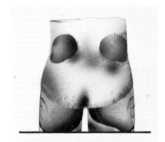

Rear View

The rear view shows that the abaxial wall at the side of the claw stands almost perpendicularly to the ground. The heel, at the back of the claw, curves slightly from the coronet to the ground. The claws are of almost equal height. The interdigital skin forms a narrow arch into which it is possible to put your little finger.

In the hind feet, the axial part of the heel of the inner claw is usually less developed than the corresponding area of the outer claw. Consequently the solear part of the heel of the inner claw has a smaller bearing surface. Furthermore, the axial weight-bearing border of the wall normally extends over a shorter distance in the inner hind claw than in the outer hind claw (see also 'bottom view').

As a result the inner hind claw has a smaller supporting surface on the axial side. It is less stable and therefore, on a hard surface, less suitable for weight-bearing than the outer hind claw.

This difference between the hind claws occurs *consistently* in many cattle breeds. In the fore claws there is no such difference; here the two claws are equally stable and are therefore equally able to bear weight.

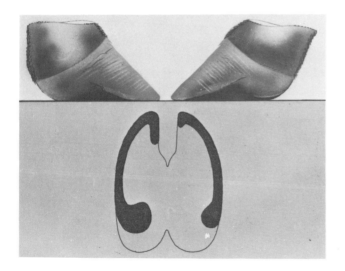

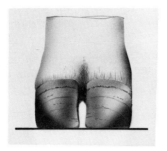

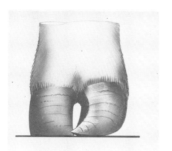

Front View

In front view – as well as in side view – the normal claw shows a more or less straight anterior margin. However, a slightly concave profile in front view should not be considered a serious abnormality (and certainly not in the fore claws), as long as the abaxial wall stands vertically on the ground. The whole of the axial wall (which is not always smooth) should also stand more or less perpendicularly to the ground.

If the walls are not upright, the claw becomes bent, forming an abnormal shape which could be called 'corkscrew claw'. In side view the anterior margin is usually concave.

This abnormality may be inborn, but it may also be the result of abnormal horn formation during the lifetime of the animal.

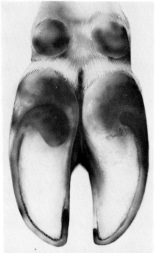

Bottom View

The consistently occurring difference between outer and inner hind claw is usually clearly visible in the bottom view as well. This is not only because the bulb (heel) and wall on the axial side of the inner claw are less developed, but also because of the slightly more concave profile of the sole of this claw.

The shape of the inner claw as a whole is somewhat different from that of the outer claw: it is more concave in the sole and less developed axially. As a consequence it rests more on the abaxial bearing border of the wall. It all boils down to the fact that the inner hind claw does not form a very stable support on a smooth, hard surface. The outer hind claw is flatter and more stable.

The fore claws are both flat and stable, and do not show any consistent difference.

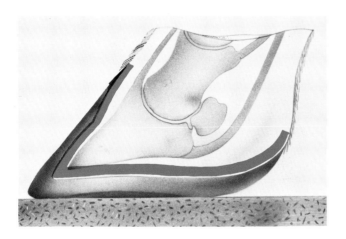

PROTECTIVE FUNCTION –
THICKNESS OF THE SOLE

The cross-section shows the inner structure and proportions of a claw, and it is apparent how thin the normal sole is. The protective function of the horny shoe is based on a sole thickness of approximately 7 mm in the area of the toe and about 5 mm in the middle. The thickness of the solear part of the heel is approximately 7 mm.

However, this thin, self-renewing layer of horn is sufficient to withstand mechanical and chemical influences from outside, and affords good protection against varying temperatures . . . within certain limits!

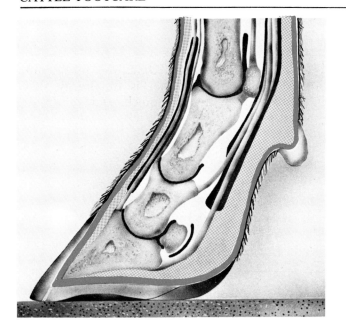

INNER STRUCTURE AND BEARING FUNCTION

In vertebrates the skeleton is the bony structure which forms the frame of the body. The skeleton in the legs supports the body.

A cross-section shows how the body weight, through **shin bone, long pastern bone** and **short pastern bone**, eventually comes to rest on the **pedal bone**.

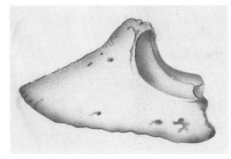

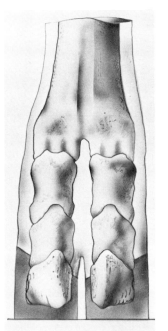

Each leg has two pedal bones, corresponding to the two digits at the end of each shin bone.

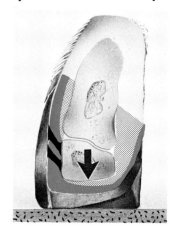

The pedal bone is surrounded by the **quick**. (This is the layer of tissue with blood vessels and nerves from which the horny shoe develops.) Therefore, the pedal bone, with the body-weight of the animal, both *rests on* and *is suspended in* this quick within the horny shoe.

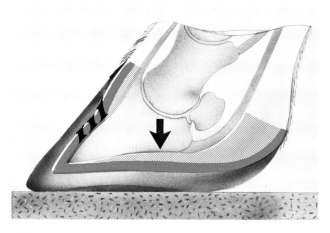

As will become apparent later in the book, the quick is the vulnerable part of the foot, and may be damaged by incorrect loading. For the moment, however, it must be assumed that the quick is not too vulnerable, and neither is the horny shoe which has to protect the quick against outside influences. If a *sound* quick had not been able to withstand the great forces within the horny shoe, and a *sound* horny shoe the great forces from outside, the cow would have become extinct before there was a chance of domesticating it.

Biomechanics

The diagrams show the influence of the body-weight on the quick in the claw, the main forces being indicated by arrows.

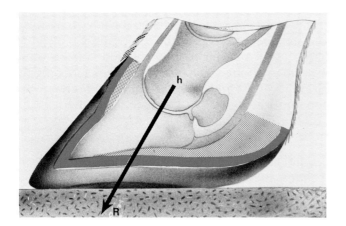

Weight is transferred to the pedal bone via the short pastern bone. The force (R), which roughly moves in the direction of the longitudinal axis of the short pastern bone, applies at the centre of rotation (h) of the pedal joint. In the pedal bone it may be resolved into a number of smaller forces, the sum of which is R, representing the load which the pedal bone exerts on the quick.

Of course, this load only exists when the horny shoe, standing on the ground, exercises an equal force on the quick but in the opposite direction.

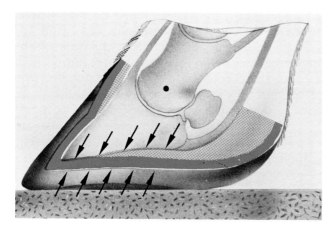

This is obvious if it is imagined that the animal is standing with one claw on an edge and the other claw hanging free. The quick in the free claw is *not* strained because there is no counter-pressure, whereas the quick in the other claw bears a double load.

The quick is under pressure between the pedal bone and the horny sole of the weight-bearing claw.

(Self-evident as this may be, it embraces the main cause of claw lameness in cattle. This consideration is also the main guide in trimming.)

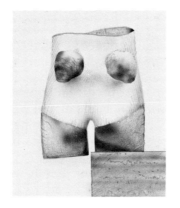

A simple calculation will indicate how much weight is carried. In a cow weighing 460 kg, approximately 260 kg rest on the fore legs and 200 kg on the hind legs. For the latter this means a weight of 100 kg per leg or 50 kg per claw.

This load must to a large extent be carried by a few square centimetres of quick under the posterior margin of the pedal bone.

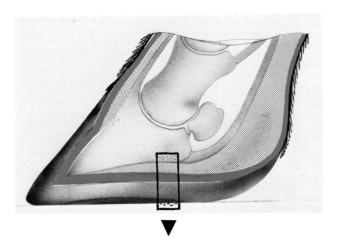

Digital Cushion – Corium – Germinal Layer

As previously mentioned, the quick in the claw is apparently well able to bear the body-weight. The claw is built for this purpose. In order to understand the effect of this weight-bearing which, nevertheless, plays an important part in claw problems a closer look must be taken at the position of the quick between the pedal bone and the horny shoe.

Surrounding the pedal bone and forming the horn, the quick connects the pedal bone with the horny shoe, as explained below (and see also page 11).

The quick consists of three layers: the **subcutaneous connective tissue**, the **corium** (or **dermis**) and the deeper layers of the **epidermis**. Connective tissue and corium contain blood vessels and nerves. Although the epidermis does not contain blood vessels and nerves, its deeper layers do consist of living cells. Therefore these also have to be regarded as part of the quick.

The corium is the strong, solid layer which isolates the body from the outside world. The epidermis is formed on the surface of the corium. In the claw this epidermis has greater thickness and strength – the horny shoe. The horn of the sole will often lose its structure on the outside, falling apart as a granular or flaky mass. Here the horn is called 'dead'. The deeper layers of the epidermis, i.e. the layers of the horny shoe towards the corium, are not directly supplied with blood, but they are living. It is here that the actual horn formation takes place, the necessary nutrients penetrating from the blood supply of the corium. The deepest layer of the epidermis, i.e. the one adjoining the corium, is called the **germinal layer**.

Often one speaks about the 'growing' of horn but in fact horn is formed like a wall is built, brick by brick. This will be discussed further on page 37.

A healthy organism is able to adapt this horn formation to what is required. Under average conditions production and wear are in balance.

Subcutaneous connective tissue makes it possible for the skin (corium and epidermis) to move. Under the pedal bone and in the heel it is amply provided with fat, thus having a *shock-absorbing function* as the **digital cushion**. In contrast with the corium and the germinal layer, the subcutaneous connective tissue does not occur in every part of the claw. In those places where it is absent, the corium is immediately adjacent to the pedal bone and is therefore *immovable*.

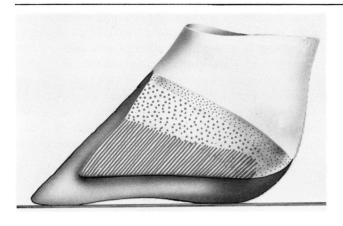

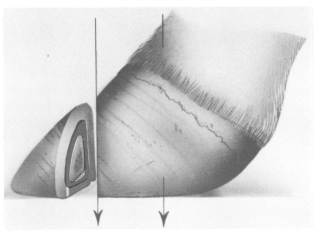

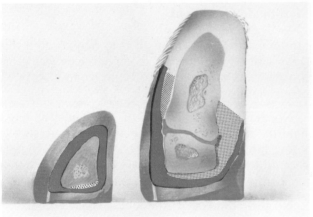

Corium and horn are, of course, immovably connected, and the strongly folded surface of the corium which penetrates into the epidermis plays a role in this.

Loading of the Quick

From the diagrams it is possible to make the following deductions: The transition of corium into epidermis (horny shoe), *in the lower half of the horny wall*, shows folds (laminae), running along the longitudinal axis of the claw. It is here that the subcutaneous connective tissue, between corium and pedal bone, is lacking. It is mainly abaxially and in the tip, and also in the anterior axial area, that the corium is firmly attached to the pedal bone. Therefore in this region the pedal bone is tightly and immovably fixed in the horny shoe.

The same applies to the solear part of the tip, where there is no digital cushion between the corium and the pedal bone.

The immovability of the pedal bone in the horny shoe is increased by the folds in the corium of the wall which make the connection between corium and horny shoe very strong.

This fixing of the pedal bone enables a firm and steady gait.

Behind the axial part of the wall, all around the heel and under the major part of the sole, the pedal bone is 'loose' due to the presence of a digital cushion. Thus there is a shock-absorbing buffer under the posterior part of the pedal bone. This enables a supple gait.

It is now possible to imagine the effect of the body-weight, as received and borne *by* the quick *through* the pedal bone:

The pedal bone *stands on* the corium of the sole and, where there is no subcutaneous connective tissue, *is suspended* by the corium *in* the horny wall (see diagrams page 20). This suspension in the vertical wall is firmest on the outside front of the claw (abaxially in the anterior half), becomes less firm abaxially and backwards, is still present to a certain extent axially in the toe, but is absent axially in the heel. It is now apparent that the body weight will cause the pedal bone to sag axially and backwards; a normal rotation as a result of its loading.

Thus it follows that pressure on the corium of the sole is mainly applied by the posterior part of the pedal bone.

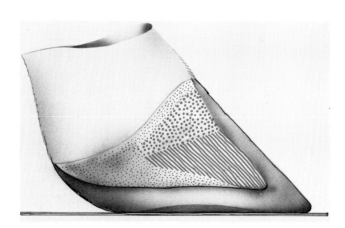

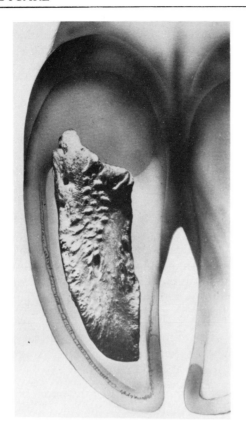

Furthermore, considering that on the axial side in the posterior margin of the pedal bone there is a thickening (the so-called **axial prominence**), it can be imagined that under heavy loading the quick under the posterior margin of the pedal bone, especially on the axial side, may become compressed. In normal weight-bearing conditions a healthy claw will be able to cope with this. But under prolonged overloading it is possible that there will be a 'vulnerable spot' in this area. . . .

This concept of the loading of the quick has been supported by the results of microscopic research carried out on the corium and helps to explain the defect which is often found in claws under the axial edge of the posterior margin of the pedal bone. For this defect the term 'typical sole-lesion' will be used hereafter.

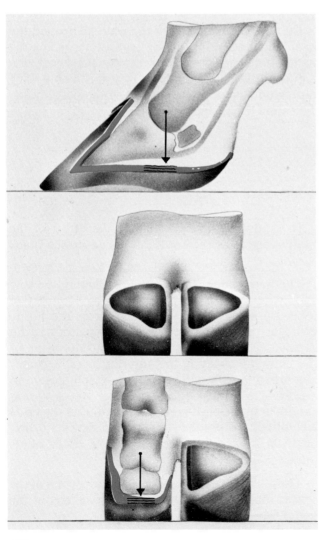

Man-made Conditions

Under natural conditions it is to be expected that cows will be moving about on surfaces which are not flat and vary in composition; sometimes hard, often yielding.

Normally the animal is in good health, has a healthy corium and germinal layer and strong, intact horny shoes. The support is stable. On hard ground the claws are worn off; on soft ground they stick into the yielding surface.

No problems occur, lameness is accidental.

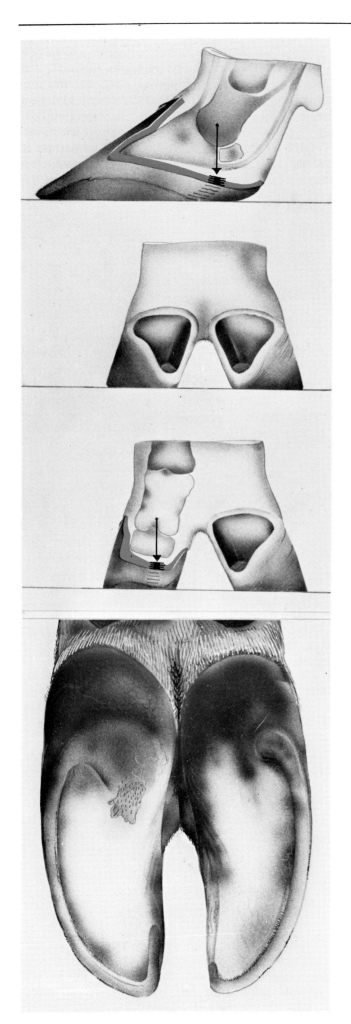

It has already been shown that under conditions of insufficient wear a claw tends to become over-grown at the abaxial wall at its anterior part (a). The hard wall develops, whereas sole and bulb show a tendency to crumble or wear off.

It will be shown below that in today's domesticated cattle as a rule, the horn of the bulb is affected and thus underdeveloped, on the axial side (b). This is a consequence of a disease which probably does not occur under natural conditions.

A long wall, abaxially in the toe, plus a heel which is axially ill-developed, result in an unstable claw which tilts over axially and backwards, at least on a hard (housing) surface (c).

Situations a, b and c are all *unnatural* to the claw; moreover they occur *systematically*.

After studying the structure of the claw the effect of these abnormal conditions can be appreciated and observed in the claw.

In a claw that leans over axially and backwards the body weight will be shifted backwards and to the axial side. The pedal bone will sag considerably, putting extra pressure on the 'vulnerable spot', which can often last for a period of six months or longer.

Although claws, that are for the best part healthy, can often stand this reasonably well, discoloration or haemorrhages and scars in the horn of the sole under the axial end of the posterior margin of the pedal bone may be regularly observed. In this case the quick is locally bruised by permanent overloading. Tissue fluid or blood will ooze from the corium, affecting the horn formation in the germinal layer and staining the horn already formed. Locally an irregular structure, like a scar, will appear in the horn. This feature, noticeable in many domesticated cattle (it is displayed by nearly all older cows), *particularly in the outer hind claw*, is referred to as **typical sole-lesion**.

The typical sole-lesion is an internal injury to the corium of the sole in the outer hind claw, as a result of unnatural conditions. The typical location is a consequence of structure and function.

Six months of housing are sufficient to reveal the vulnerable spot; however this process does not necessarily result in serious trouble. Without any additional diseases it will not usually go beyond a stain or a scar in the horn.

A cow may be tender-footed but there is no real lameness problem.

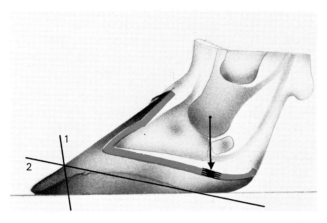

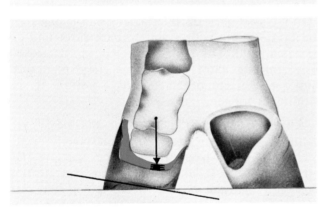

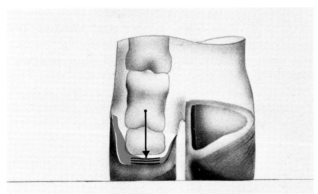

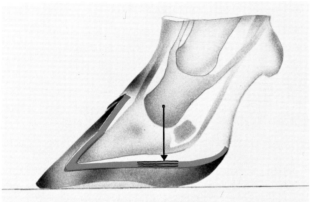

It is now possible to understand the effect of shortening a claw (1), and of cutting or trimming a long abaxial wall (2). The claw tilts forwards and outwards (forwards and abaxially), so that the body-weight is shifted in that direction. The claw will rest more on the wall, where a strong vertical connection helps to absorb the force of the body-weight. Thus the quick in the sole is loaded more evenly and, at the same time, less heavily. The same would apply to a long wall on soft soil; the long nail is pushed down and the claw stands upright.

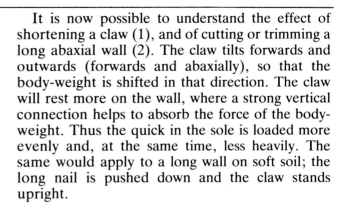

Healthy claws under natural conditions can do their job without any special care. It is when housing conditions cause insufficient wear that claw trimming becomes desirable or necessary.

It will now be appreciated why a wall has to be kept short in order to make it bear.

In spite of good care, however, hard floors are not really suitable for a prolonged stay. In this respect the difference between the claws plays a part.

DIFFERENCES BETWEEN THE CLAWS

It is generally known that it is particularly the *outer hind claws* which are a source of problems. It is on the outer hind claws that typical sole-lesions systematically occur and it is there that lameness usually affects the dairy herd.

The inner hind claws and the front claws have so far not been a real source of anxiety.

Why are the outer hind claws more susceptible to problems?

The answer is related to two factors:
- the structure and function of the foot;
- the diseases of the foot.

Further points about function and structure will be dealt with in this chapter while the pathological aspect will be discussed in the following chapter.

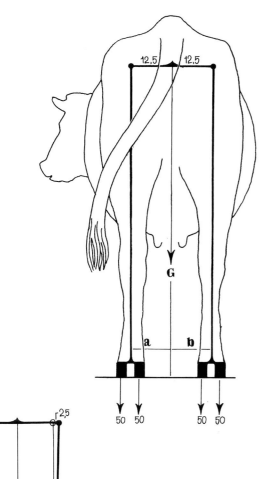

Biomechanics

The distribution of weight over the two claws of one foot depends, among other factors, on whether the claw is situated on the *outer* or the *inner side*. Biomechanics applied to the hind legs of cattle indicate the following points.

Hind legs are connected with the pelvis by a ball-and-socket joint. In a 'squarely' standing animal the part of the body-weight borne by the hind legs (G), will be evenly distributed over the two legs and, if the claws are equally high and stable, over all four claws:

outer left	inner left	inner right	outer right
50kg	50kg	50kg	50kg

Convenient and realistic figures have been chosen for the width of the pelvis at the place of the hip-joints and for the weight resting on the hind legs.

In the ever-present possibility of small movements sideways, i.e. small sideways deviations from the square position, the following situation could arise.

The weight, distributed over the two legs according to this formula:

$$\text{weight left} = G \times \frac{b}{a+b} \quad \text{and weight right} = G \times \frac{a}{a+b}$$

will be redistributed as follows:

$$\text{weight left} = 200 \times \frac{10}{25} \quad \text{and} \quad \text{weight right} = 200 \times \frac{15}{25}$$

$$= 80 \text{ kg} \qquad\qquad\qquad = 120 \text{ kg}.$$

As might be expected, the leg is more heavily loaded on the side to which the cow leans over. It is perhaps surprising that such a small movement should create such a large difference.

27

If the two claws were attached to the leg in a *perfectly elastic* way, the weight, thus distributed over the two legs, would be evenly sub-divided over the claws of each foot:

outer left	inner left	inner right	outer right
40 kg	40 kg	60 kg	60 kg

If the connection of the two claws with the leg was *quite inflexible*, the new weight would rest fully on the claws at the side to which the animal leans over:

outer left	inner left	inner right	outer right
0 kg	80 kg	0 kg	120 kg

In reality, what happens lies somewhere in between, for the claws are connected with the leg in a stiffly elastic way. The distribution of the new weight over the two claws of the foot will be somewhere between 50%–50% and 0%–100%; for instance, 40%–60%.

This means that three-fifths of the new weight is borne by the claw on the side to which the cow tilts over, and two-fifths by the other claw.

A simple calculation shows the new loading of the claws:

outer left	inner left	inner right	outer right
2/5 × 80	3/5 × 80	2/5 × 120	3/5 × 120
= 32 kg	= 48 kg	= 48 kg	= 72 kg

In case of a similar movement towards the other side:

72	48	48	32

In other words, the **weight-bearing** by the outer hind claws **varies widely** with the successive movements of the body. The load on the inner hind claws is far more even.

The ever-shifting distribution of weight over the two hind legs is constantly 'corrected' by the outer claws. Hard surfaces (housing) will enhance this effect.

N.B. This does not apply to the forelegs. Instead of being inflexibly connected with the body by a joint, their connection is cushioned by the tendons and ligaments of the shoulder blades.

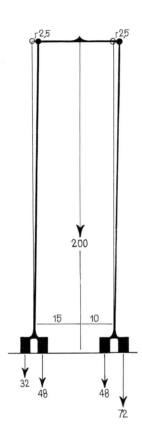

Functional Anatomy

Functional anatomy is the science concerned with the relationship between structure and function. A 'rough' surface on a bone indicates regular forces exerted on the bone by the attached tissue. Correspondingly this tissue will itself be regularly submitted to the same forces.

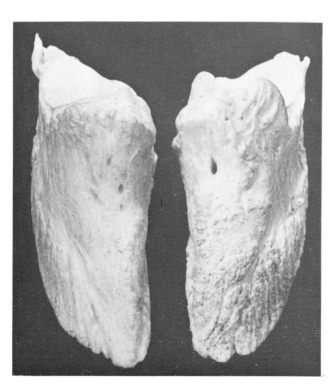

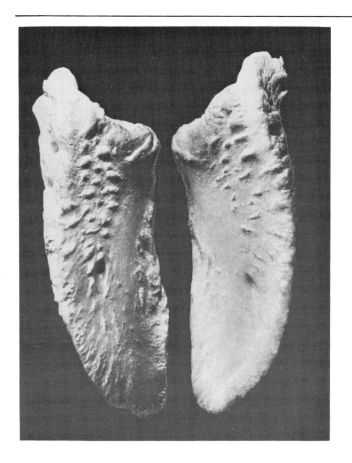

These tissues consist not only of tendons and ligaments, but also of the quick where it is attached to the pedal bone.

The 'smoother' the surface of the bone, the smaller the forces, the 'rougher' the surface, the greater the forces.

An examination of the pedal bones of hind claws shows that those of the outer claws are always rougher than those of the inner. In young animals this difference is minimal, but it increases with age as the forces exert their influence over a longer period. Then the pedal bones of the outer hind claws grow increasingly rough, which indicates that the forces affecting the outer claws are greater than those affecting the inner ones.

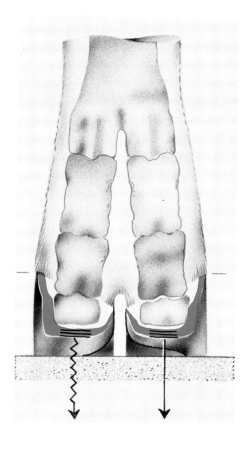

This finding supports the biomechanical calculations that the outer hind claw is more heavily stressed than the inner. 'More heavily' in this case has a *qualifying* sense: more irregularly which means unfavourably. For, according to the rule of functional anatomy, an irregular and ever-shifting weight is difficult to bear. Such an irregular load on the pedal bone of the outer hind claw must be very 'tiring' for the quick in this claw. The corium will easily become 'over-exerted' and irritated. Such a condition is often accompanied by an increased tissue activity.

In the forelegs there is no such obvious systematic difference in roughness between the pedal bones.

Enlarged Outer Hind Claw

Irritation of tissue may cause excessive growth (so-called hypertrophy), as well as an increased cell-division (hyperplasia). These two processes may be regularly observed in the outer hind claws of our adult cattle under the prevalent farm conditions: hypertrophy of the corium and hyperplasia of the germinal layer resulting in an **enlarged outer hind claw** showing a visibly larger size, a wall grown higher, and the horn of the bulb and sole grown thicker.

29

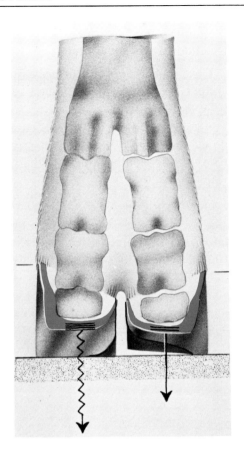

The higher claw will start to bear a greater part of the body-weight, at least on a hard surface. In so doing the outer hind claw, which is unfavourably loaded, will after some time become *over*loaded as well. This will again lead to irritation connected with hypertrophy and hyperplasia. . . . A **vicious circle** is liable to arise, causing the outer hind claw to suffer increasingly.

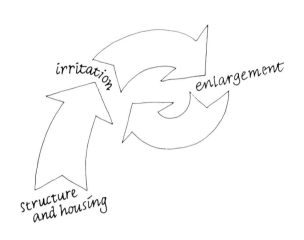

irritation

enlargement

structure and housing

N.B. It seems reasonable to expect that the extra height produced by an increased weight-bearing would be counterbalanced by extra wear, until the point is reached where height – and thus weight – are approximately equal. This however, is not the case; the higher claw remains higher and more heavily loaded. In two claws of different heights the equilibria between growth and wear appear to be situated at different 'levels'.

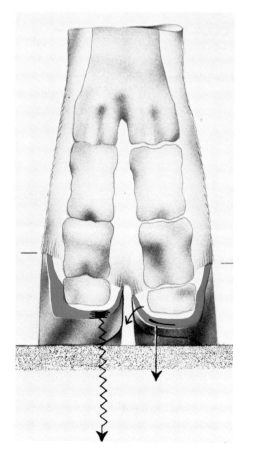

An additional cause for *an even greater load* on the outer hind claw is the shape of the inner hind claw. Being less stable this claw is liable to heel over axially under its weight-bearing, certainly on hard surfaces. Part of the weight is so to speak transferred to the outer hind claw. This means an extra load, especially axially, on a claw that is already at a disadvantage because of its position on the outside of the leg.

ulceration

Contusion
(sole lesion)

aberrant
horn formation

excessive
horn formation

structure
and housing

Typical Sole-Lesion

The **typical sole-lesion** is a contusion of the quick, involving the formation of a scar or a discoloration in the horn, at the 'vulnerable spot' in the outer hind claw. This weak spot is also referred to as the 'typical place'. In cases of prolonged housing on a hard surface, the typical place in the overloaded claw cannot bear the body-weight without injury.

The typical sole-lesion is a consequence of the build of the foot and the unequal weight distribution connected with it. It is brought about by housing conditions which systematically enhance tilting over as well as over-loading.

Therefore the typical sole-lesion is not a disease, but a symptom of domestication.

It is seldom found in the inner hind claws, where weight-bearing is more even and less weight is borne due to the difference in height. In forelegs typical sole-lesions are occasionally found in the inner claws, which are often somewhat higher than the outer ones. Apart from this, these claws have been very little studied.

The vicious circle that is always liable to develop in the outer hind claw may, through a typical sole-lesion, produce defects in the horn of the sole known as **sole ulcers**. These will be discussed in the next chapter.

It should be noted here and now that **under the prevalent conditions a healthy outer hind claw is able to function for a long time, if reasonable care is observed!** . . . albeit not without getting scars.

In a healthy organism the consequences of the build and function of the hind feet, in which housing plays an important role, remain limited. **In a healthy quick, hypertrophy as well as hyperplasia are greatly restricted**, certainly if regular claw care prevents prolonged overloading of the outer hind claw:

- Cutting off walls which are too long, to achieve a favourable, upright position of the claws and to improve the stability of the inner claw.
- Trimming overgrown claws, to equalise the quantitative weight distribution between two adjacent claws.

These matters are discussed extensively under Chapter Three, 'Trimming'.

Measuring Weight Bearing

The problems of weight bearing in the hind foot indicated by the discussion of biomechanics and functional anatomy have been confirmed by simple measurements, showing the quantitative as well as the qualitative results.

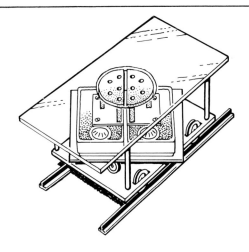

The measuring equipment consists of two platform scales, taking up to 120 kg maximum each and placed next to each other. The unit is assembled behind the floor of a crush in a movable way so that it can be adjusted to the stance of the animal. A cow is led into the crush so that the inner and outer claw of one leg each rest on one scale. The scales will then indicate the weight distribution between the two claws, and the fluctuation of weight on each claw caused by slight movements of the animal. With a little patience this is a fairly simple procedure, and the basic posture of the animal is unaffected.

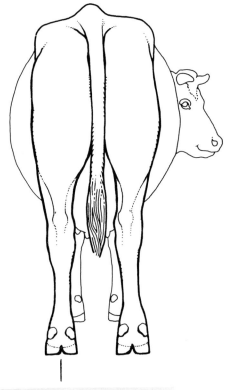

POSTURE OF THE HIND LEGS

There is a connection between the posture of the hind legs in rear view, and the condition of the claws.

Normally the hind legs are more or less parallel. It is hardly surprising that in the course of a few years the difficult task of the outer hind claws may affect the posture of the hind legs.

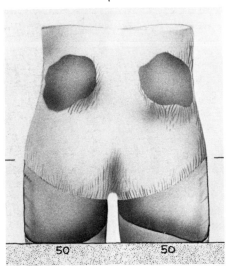

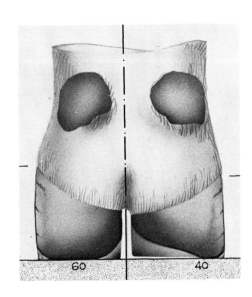

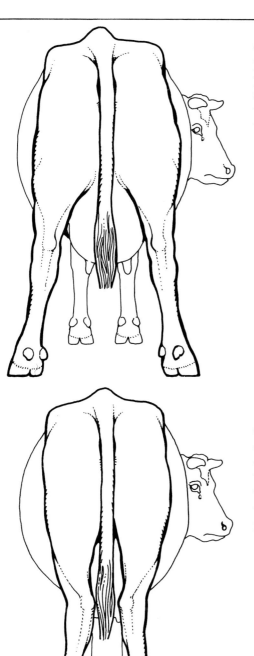

Heavy loading and contusion of the quick in the outer hind claw will in the end cause sensitivity or pain. The cow will try to compensate for this by placing or turning the foot more outwards (by adopting a base-wide or a cow-hocked posture), thus making the inner claw bear more weight and the outer less.

Such an adjustment of posture is frequently observed. Thus the posture of the hind legs of our dairy cattle is only partly *inborn*, and to a great extent *adopted*. The latter is due partly to the unfavourable loading on the outer hind claw and to domestication, but probably for the greater part to the claw diseases that will be discussed in the next chapter.

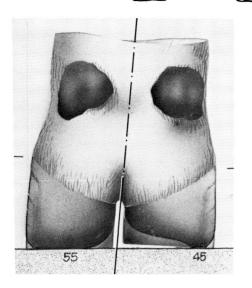

As a matter of fact, the effect of this adjustment of posture is relative: measuring shows that in adopting a cow-hocked or base-wide posture, the cow relieves her sensitive outer claw of its over-load *only to a certain extent*. Mostly this claw remains more heavily loaded than the inner claw, albeit less than before. This must be ascribed to the difference in height.

One aim of trimming is to improve the weight distribution in the foot. If this succeeds many animals will resume a normal, or more normal, stance.

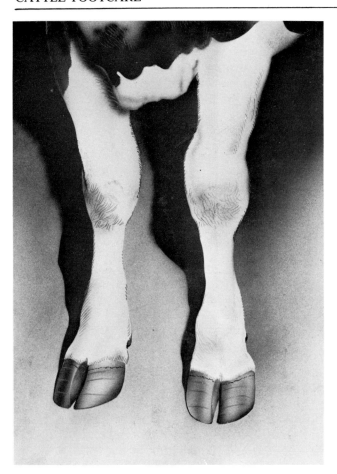

FORE CLAWS

Why do the claws on the forelegs have so few problems?

Firstly they are unlikely to experience the systematic biomechanical differences that are so striking in the hind claws. The fore claws are far more equal and are both quite stable.

Secondly, it is possible that they remain much *healthier* than the outer hind claws.

The pathological aspect of the lameness problem in cattle where functional anatomy, as described above, plays an important part, will be discussed in the next chapter.

N.B. Recently it has been found that cattle housed in cubicles, particularly those inside the whole year, can be subject to inadequate horn formation in the front claws (pages 51 and 120).

Chapter 2

DISEASES OF THE QUICK

Dairy cattle are kept for milk production . . .

A good yield for a prolonged period of time can be expected only from a *healthy* cow. A healthy cow that has to serve a good many years in modern dairy farming must have healthy legs and claws.

A healthy claw means a healthy quick. In this chapter, the term 'quick' refers in particular to the corium and germinal layer, as it is here that the formation of the horny shoe takes place. This quick must not be unhealthy, for only healthy tissue and good horn production will serve the twofold task of bearing weight and offering protection.

Moreover, adaptation to changing circumstances also requires a healthy quick.

DISEASE AS A HERD PROBLEM

Two diseases of the quick in the claw are known to occur *often and almost everywhere* in highly productive animals in intensive dairy farming. These are 'heelhorn erosion' (also known as 'slurry heel' or **interdigital dermatitis**) and **laminitis**.

The diseases often occur together, the symptoms of both influencing the outward appearance of the claw. This sometimes makes a positive recognition rather difficult.

Factors like breed, climate, housing, food and care all influence their occurrence and severity to some extent. In addition to the direct cause and the hereditary disposition (susceptibility), it is quite understandable that living conditions, too, may play an important role in the origin of diseases. For example, this applies to bacterial disorders of the foot: under humid and warm conditions bacteria may thrive and spread in soiled bedding, thus increasing the chance of contamination and infection, but will probably not survive in very dry or cold surroundings. As for metabolic disorders: it becomes more and more evident that wholesome, and at the same time economic, feeding becomes more difficult as productivity increases; the farmer is therefore an important factor.

In Holland both diseases are widespread and often serious among the Dutch Friesian breed **in all types of farming systems**.

Interdigital dermatitis and laminitis regularly occur as a **herd problem**, but it should be noted that, fortunately, not all farms are affected by these diseases and, if they are, not to the same extent. The problem, however, is an important one, and high production and intensive farming seem to be involved. With this in mind dairy farmers in future will have to pay a good deal of attention to claw diseases.

Although no figures are available, it is widely supposed that the M.R.IJ. breed (Dutch red-and-white) are less affected. The same symptoms may be found in this breed, but they seem to be less widespread and not as serious. Interdigital dermatitis and laminitis also occur in dairy breeds outside Holland, albeit in varying degrees. Both hereditary disposition and external circumstances play a part in this. The Dutch Friesian breed is generally known to be susceptible to claw troubles whereas the American Holstein Friesian *is reputed* to be less susceptible.

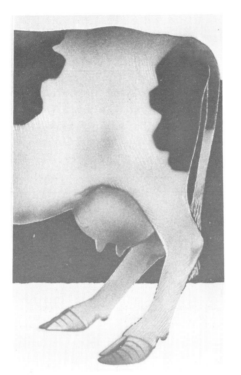

At the beginning of this chapter it was stated that a cow must have healthy legs and claws. Only *foot* diseases and *foot* problems will be dealt with here; disorders of the *legs* will not be discussed. In at least 95 per cent of the cases, however, lameness and tenderness are the result of conditions of the foot.

Tarsal bursitis or 'concrete hock' is in fact the only condition of the leg which sometimes forms a herd problem, at least in the tie-up stall. The frequent occurrence of 'weak' stances of the hind legs is nearly always the result of diseased claws, with abnormal horn formation.

The problem of lameness in dairy farming is a problem of **claw lameness**.

STRUCTURE OF THE SKIN

It is useful here to go further into the structure of the skin and formation of the epidermis, in order to gain a better insight into the progress of claw diseases.

The body is completely covered with the **skin**. The **epidermis** of the skin forms the actual protective outer layer. Like human nails, a cow's horny shoe is part of the epidermis.

The epidermis contains no blood vessels or nerves. The outer layer peels or wears off, and could be called 'dead'. The inner layer of the epidermis is called the **germinal layer**, and is definitely *not* dead. Like the next few layers of the epidermis, this layer must be considered part of the 'quick'. It has, however, no *direct* blood supply and no feeling.

The 'real' living layer of the skin, the **corium** (or dermis), lies inside the epidermis. It forms the outermost organ of the body containing blood vessels and nerves. Damage to the epidermis does not hurt and does not cause any haemorrhage; damage to the corium, however, does cause both pain and haemorrhage.

The partition (or connection if you like) between corium and germinal layer is called the **basement membrane**. Through it, from the blood vessels in the corium, nutrients can reach the germinal layer.

Horn Formation

Horn formation has its origin in the *germinal layer*.

The epidermis grows thicker as a result of cell production (by a continuous division of cells) in the germinal layer; newly produced cells push the previously formed cells towards the surface. The nutrition needed for this cell division originates with the blood supply in the corium and diffuses through the basement membrane. The shifting cells produce the horny substance at their own surface, but when they have moved further from the basement membrane they no longer receive nutrition. In the end they have completely replaced themselves by horn; the epidermis has become *keratinised*.

Thus the epidermis generates itself by cell division in the germinal layer, but it does so with the help of nutrients from the corium. Like a healthy germinal layer, a healthy corium is indispensable for the formation of a sound horny shoe.

The relationship between corium and germinal layer is a very close one. They mutually influence each other.

A disorder of the corium will have its effects on the cell formation in the germinal layer and thus on the generation of horn. Horn formation will often be stimulated by a more intensive blood circulation as a result of this disorder. In certain places, however, the production of horn may be disturbed by a local obstruction in the blood vessels, e.g. as a result of continuous bruising, or clotting of blood.

On the other hand the corium will *react* to a disorder in the germinal layer; for instance by means of a better blood circulation, indicating an increased activity. Horn formation thus may stagnate *in a certain area* because of a localised disorder in the germinal layer, but at the same time may be stimulated *in the rest of the claw* by the general reaction of the corium.

A diseased quick forms either too much or too little horn . . . and besides the horn is often of poor quality. The balance between supply and demand is disturbed.

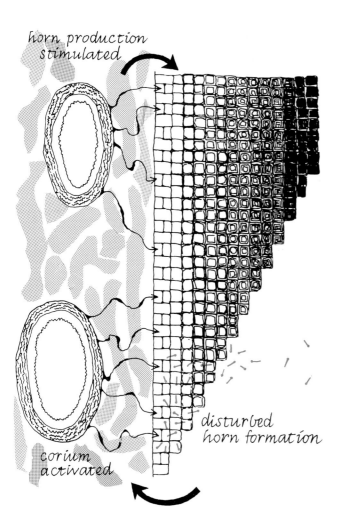

horn production stimulated

disturbed horn formation

corium activated

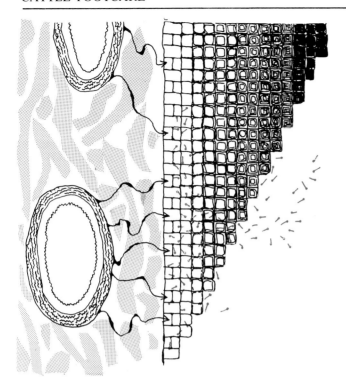

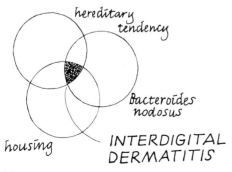

hereditary tendency

Bacteroides nodosus

housing

INTERDIGITAL DERMATITIS

INTERDIGITAL DERMATITIS

As far as we are aware interdigital dermatitis (slurry heel, or heelhorn erosion) is a bacterial disease of the epidermis of the interdigital skin and the bulb of the claw. The generation of new cells in the germinal layer, and the formation of horn by these cells, are at risk. Eventually the corium will be irritated by the disorder. However, the bacteria do not enter the corium as they do not pass the basement membrane.

Bacteroides Nodosus

For the present *Bacteroides nodosus*, a rod-shaped bacterium with a knob at one end, has been identified as the cause of the condition. Almost everything we know about its life cycle has been learnt from the study of **footrot** in sheep. This contagious claw disease in sheep is a condition similar to interdigital dermatitis in cattle. Wherever in the world intensive sheep or cattle farming are practised, footrot or interdigital dermatitis will appear.

A high concentration of cattle makes the occurrence of the disease more likely, due to the abundance of infectious material left in the ground cover (straw, dung) by the infected animals.

A warm humid environment weakens the skin, thereby making penetration into the epidermis easier for the bacteria. In addition, these conditions activate the agents of the disease. Under cold or dry conditions they are considerably less active.

Fusobacterium necrophorum, the bacterium that produces foul-in-the-foot (see below in this chapter), is thought to assist *Bacteroides nodosus* in penetrating into the epidermis. *Fusobacterium necrophorum* is one of the normal inhabitants of the digestive tract in cattle and is therefore always found in the environment.

Bacteroides nodosus is the specific agent of the disease. It may survive in the epidermis of cattle and sheep. Through these animals the ground cover may be contaminated as well, but it is generally assumed that *Bacteroides nodosus* can only survive outside the host for about 2–4 weeks.

If it is also taken into consideration that susceptibility varies between breeds and between individuals in each breed, it can be seen that a *multifactorial disease* is being dealt with here. That is a condition in which several factors *combined* give rise to a certain characteristic pattern of symptoms (a syndrome). (See also page 107 et seq.)

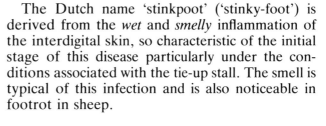

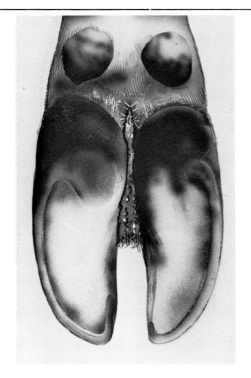

The Dutch name 'stinkpoot' ('stinky-foot') is derived from the *wet* and *smelly* inflammation of the interdigital skin, so characteristic of the initial stage of this disease particularly under the conditions associated with the tie-up stall. The smell is typical of this infection and is also noticeable in footrot in sheep.

In the following pages the term 'heelhorn erosion' will be used for everything resulting from an invasion of *Bacteroides nodosus* in the epidermis. When the symptom of the wet inflammation of the interdigital skin becomes less noticeable during the progress of the disease or under certain housing conditions, heelhorn erosion remains the most striking symptom.

The name 'footrot', used to describe the well-known foot problem in sheep, should not be used in connection with cattle, to avoid confusions caused by using the same term for different clinical pictures.

Two Stages

Heelhorn erosion entails several symptoms, in which we must recognise two stages in order to arrive at a clear understanding of the course of the disease.

In **stage 1** the epidermis is affected by the specific agent (the *causal* infection), which has an effect on its formation. The noticeable consequences are:

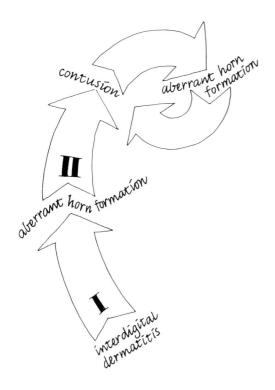

- inflammation of the interdigital skin,
- destruction and defective formation of the horn of the bulb, and
- excessive horn formation beyond this affected bulb area.

Stage II reveals the effect of this horn destruction and abnormal horn formation on the support and loading of the corium; and the consequences of these effects, again for the corium and the horn formation. Furthermore, defective horn formation gives numerous types of bacteria in the environment a chance to penetrate, giving rise to *secondary* infections.

Once 'heelhorn erosion' has brought on major changes to the horny shoe, these will remain, even if changing circumstances make the causative infection disappear. So stage II can simply continue, regardless of the original infection; then stage II has taken its own, independent course.

It is necessary to have a clear insight into the two stages of interdigital dermatitis, to be able to take the right measures at the right time when fighting this disease and its consequences.

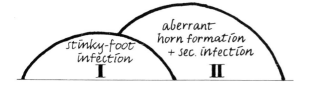

39

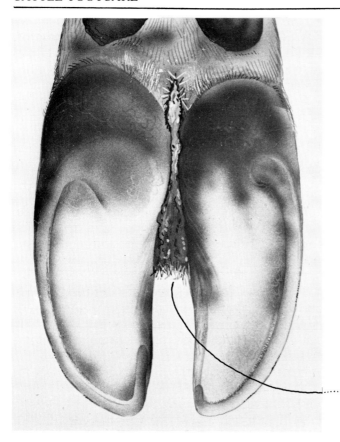

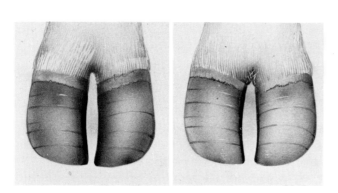

Stage I

a. Inflammation of the interdigital skin

The most typical initial symptom of heelhorn erosion is a wet inflammation of the interdigital skin. The inflammatory fluid, translucent at first, soon becomes grey-coloured. This is a result of disintegration of the epidermal cells and liquefaction of the horny substance due to the activities of the causative bacteria.

The corium is never subject to such disintegration. In a smear for microscopic examination, in which material is made visible by staining, no inflammatory cells (which would have originated in the corium) are to be found; nor is the inflammatory fluid purulent or yellow. The basement membrane remains intact; the causative infection does not penetrate into the corium (diagram, page 38).

The corium, however, does show signs of increased activity, such as more intensive blood circulation and a concentration of cells on the side of the epidermis. This reaction of the corium to the bacterial infection of the deeper layers of the epidermis often results in a swelling of the interdigital skin plus underlying tissue. The interdigital skin is as it were pressed together, one side to another.

The inflamed interdigital skin feels wet and sticky and is sensitive to the touch. The inflammation ends where the hairy skin begins, both in front and at the back of the interdigital space.

The animals are not lame, or hardly so, although they are occasionally somewhat uneasy.

As mentioned above, heelhorn erosion is a multifactorial disease. Housing is one of the factors that strongly influence the seriousness and course of the disease. The wet interdigital dermatitis will be particularly apparent where cows are tethered indoors with straw-covered floor, during the first few weeks or months of the housing period. There it occurs predominantly in the hind feet, warm wet straw favouring the conditions. The forelegs are much less affected, perhaps because the straw is dry.

In a number of animals one or more legs may become affected. It is striking that the symptoms may occur almost simultaneously in different parts of the cowhouse, whereas nothing had been noticed in the pasture shortly before. The animals must have been contaminated already, the bacteria waiting only for a change in external conditions.

The inflammation of the interdigital skin is far less distinct in the cubicle system. The overall damp and dirty cover of the ground obscures the clinical picture, while at the same time the temperature between the claws may be lower. The disease takes a less virulent course, but is more widespread, affecting more animals and, in particular, more forelegs.

When the cows are turned out to grass the inflammation of the interdigital skin often spontaneously and quickly disappears; at least clinically, which means as far as is visible. (In extensive cattle farming in a hot and dry climate, little heelhorn erosion will occur.)

There is a wide variation in the seriousness and duration of the first stage of heelhorn erosion, factors such as housing, ground cover and indoor climate all playing their parts. Sooner or later, the infection will spread to the bulb of the claw.

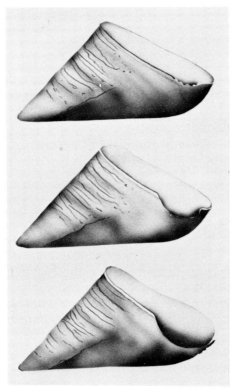

b. Defective horn formation in the bulb area

The invasion of *Bacteroides nodosus* is apparently able to spread from the interdigital skin in the direction of the bulb of the adjacent claws. Where the germinal layer and the horn-producing cells are affected, the connection between the corium and the horn of the bulb may become locally disrupted (diagram p. 38). This 'heelhorn erosion' will **undermine** the **horn of the bulb**, first on the axial side and then further abaxially in the direction of the toe (compare page 55, 2nd par.).

The process may be slow or fast, either with or without production of new horn. The result will be the appearance in the bulb horn of **fissures** of varying depth, always running from axial backwards to abaxial.

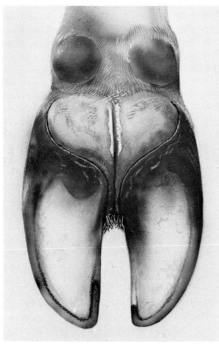

Any new horn in the undermined area is of poor quality and often lacks a solid structure. Depending on this horn formation, the bulb behind the fissure will be scarcely or only poorly covered with horn.

The appearance and the degree of the defective horn formation in the bulb are to a large extent determined by housing, ground cover and indoor climate.

The destruction of the horn is confined to the bulb of the claw. In an advanced case, therefore, a fissure may reach as far as the most posterior part of the side of the claw.

In the second half of the housing period horn deformation becomes more conspicuous, whereas the inflammation of the interdigital skin itself often becomes less manifest.

During the grazing season horn formation in the bulb is readily and spontaneously restored. The old deformities, of course, will take some time to 'grow off'. Trimming may be an important aid here and considerably accelerate the process of restoration.

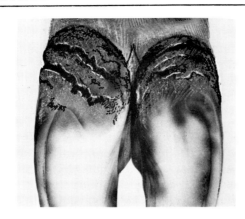

The formation of fissures generally is less conspicuous in cubicle housing, the 'heelhorn erosion' instead tends to be composed of irregular ridges. This condition is also referred to as 'slurry heel'.

If the housing is used during the whole year, deformation of bulb horn is noticeable throughout the year.

The process described in section b is an alteration of the horn as a consequence of the process described in section a. A fissure is a horn deformation following on from the inflammation of the interdigital skin.

As a result of pressure by the edge of horn on the bulb-tissue behind, *fissures may lead to obvious lameness* (stage II).

c. Excessive horn formation

At the beginning of this chapter, under 'horn formation', it was stated that a disease of the epidermis can activate the corium. In addition to a local stagnation, as described under section b, heelhorn erosion may cause horn production to increase in the rest of the claw.

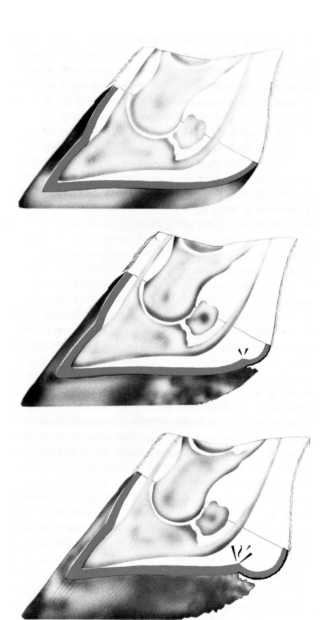

Depending on the degree of infection on the farm, heelhorn erosion will systematically cause the formation of *too much* horn; the claws 'grow' too fast. This is the excess horn which is to be cut off in trimming.

The claws grow longer and, also, *higher*. In serious herd infections this phenomenon occurs in all claws, but *as a rule particularly in the outer hind claws*. Under abnormal conditions, in the case of illness, the outer hind claws, unfavourably loaded, will be the first to become irritated and react excessively. Horn formation and wear will no longer be in balance.

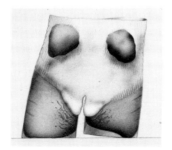

So during the housing period there will not only be an increasing number of fissures, but also more and more *outer hind claws that are too high*. These grow considerably higher than the adjacent inner claws. During the summer grazing this imbalance in horn production will decrease because interdigital dermatitis will heal.

Where cubicles are also used for summer housing heelhorn erosion is more permanent; unsuitable growth of horn occurs more or less continuously. . . .

The process described in section c is an alteration of the horn, as a result of the processes described in sections a and b. Excessive horn formation under the outer hind claw is a sequel to the inflammation of the interdigital skin and the destruction of the bulb horn.

Because it overloads the outer hind claw, *excessive horn formation may lead to obvious lameness* (stage II).

Stage II

The lamenesses mentioned under sections b and c indicate *pain caused by pressure*. They appear during the second stage of the heelhorn erosion syndrome, and are the result of the effects on the quick in the claw caused by the horn deformities produced in the first stage of the disease.

Stage II concerns the **inflammation of the quick which results from mechanical damage caused by incorrect horn production**; the secondary traumatic disorders of the corium and digital cushion.

Lameness as part of the heelhorn erosion syndrome is always due to secondary and traumatic inflammations. The primary disease of the epidermis (the horn-producing cells being affected by *Bacteroides nodosus*) does not cause more than a slight tenderness of the skin at the most, but no real lameness. This is in contrast to the situation in laminitis.

The effects on the quick produced by the deformities of the horny shoe can be reduced by cutting off these deformities, i.e. by trimming the claw.

Events in the Second Stage

The seriousness of the events in the second stage depends on the severity of the causal infection (stage I) underlying it. Severe infections are accompanied by extensive horn changes which may deeply damage the quick. Slight changes of horn resulting from a mild infection will damage the quick to a much lesser extent.

Once the second stage has begun, however, unsuitable horn formation and damage to the quick may keep each other going, even without any further stimulus of an active causal infection. The second stage is ready to be self-sustaining.

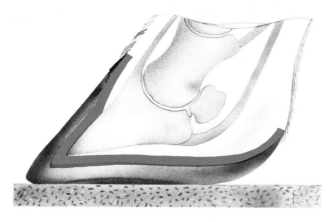

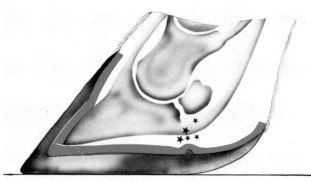

In Chapter One ('Structure and Functions') it was emphasised that in the weight-bearing claw forces are transferred to the corium of the sole from the pedal bone. It was also stated that in this case the horny sole exercises equal but opposite forces on the corium.

The corium occupies a position between the pedal bone and the sole and is smoothly and evenly supported, the strongest pressure on it being exerted near the posterior margin of the pedal bone (mainly axially). In other words, although pressures on different parts of the corium may not be the same, the support is smooth and even. The cow stands with its 'quick' on a smooth 'floor'.

If the borderline between the corium and the horny sole is not smooth, due to a lump or ridge somewhere in the horn, then *extra counter-pressure* will be exerted on the corium in that place. This uneven pressure will in the end contuse the quick, especially the corium. The same occurs if there is a depression in the horny sole, or if it ends abruptly, part of the corium being no longer supported. Sooner or later, the corium will be locally damaged by **the pressure of the edges of the horn**.

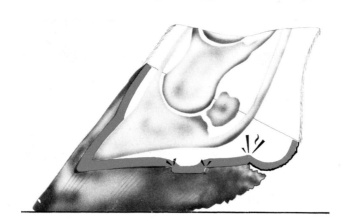

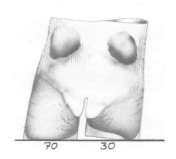

The load on the quick in a claw depends, among other things, on the *difference* in height between the two adjacent claws. The higher claw bears the greater part of the weight and there will be greater forces on the quick in this claw.

Correspondingly the corium of the higher claw experiences greater counter-pressures than there are on the corium of the lower claw which bears less weight. If there are any irregularities on the inner surface of the horn their effects on the corium will be greater in the higher claw. There will be less effect on the lower claw: less weight-bearing, less counter-pressure.

The course of the heelhorn erosion syndrome will now be considered in some detail.

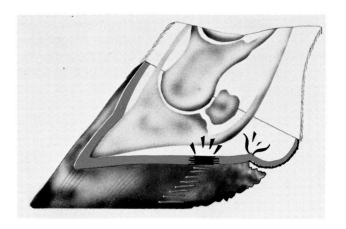

The defective horn formation in the bulb gives rise to ridges and fissures. The even support of the corium will be broken to an extent which depends on the depth of these deformities. *The horny edge of the fissure* (the hard anterior margin!) *presses into the corium.* This is the case behind the posterior margin of the pedal bone, in the area of the solear part of the bulb. It may happen both to the inner and to the outer claw, the extent of pressure depending on the depth of the fissure. The often considerable swelling of the bulb area behind the fissure will be the result of this pressure.

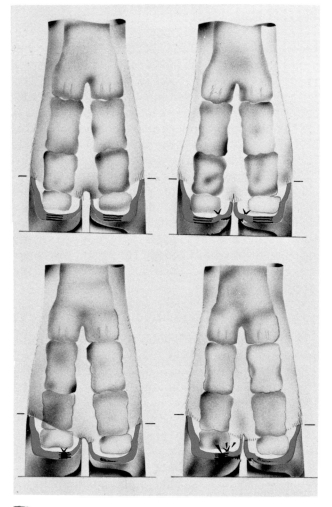

In the course of the heelhorn erosion syndrome, however, the outer hind claw grows *considerably higher* than the inner hind claw. With the exception of the posterior part of the bulb, where stagnation of horn formation leads to loss of horn, excessive horn formation causes the wall to grow longer and higher and the sole to grow thicker. **The outer hind claw becomes overloaded!**

The consequences for the corium are clearly visible. The corium is contused by the horny edge originating in the depth of a deeper fissure, but especially in the excessively high outer hind claw. In the inner claw which bears less weight the corium is subjected to far less pressure.

In the higher outer hind claw, which sometimes has to bear as much as 70 per cent or more of the hind leg's weight, the corium will also be subject to heavy pressure under the posterior margin of the pedal bone, particularly on the axial side (the 'typical place'). Here, as it were, a **typical sole-lesion** appears which is **aggravated** by excessive overloading.

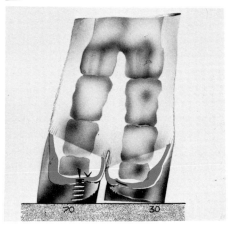

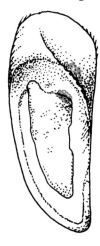

N.B. In the chapter on 'structure and function' weight-bearing in a healthy foot was discussed. Although some theoretical problems were pointed out, in practice these do not cause many difficulties. In the present chapter, however, dealing with **claw lameness as a herd problem**, the discussion is no longer about pressure on a healthy quick, but **incorrect pressure** on a **diseased quick**. It is this very *combination*, partly springing from the specific pattern of weight-bearing in the hind foot, that gives rise to the problems of the outer hind claw.

Vicious Circle

Abnormal horn formation leads to contusion, but due to the reaction of the corium, *contusion* again *results in abnormal horn formation*. A vicious circle develops and continues, produced by problems of load and support in the corium. Stage II has taken its own independent course as part of the heelhorn erosion syndrome.

Prolonged overloading causes contusion. Contusion primarily causes the excretion of tissue fluid and the occurrence of haemorrhages in the corium. In the end insufficient local blood circulation (ischaemia through stricture of blood vessels and thrombosis) will reduce the tissue resistance and eventually lead to a local death of tissue (necrosis).

This process reveals itself as follows:

If tissue fluid and blood pigment reach the epidermis through the basement membrane, yellow or red discolorations in the out-growing horn (a 'bruising') will show evidence of contusion.

Later, if the blood supply in the area of contusion (often the typical place) is obstructed, horn formation may stagnate locally. A concavity will be formed on the inside of the horny sole; that is, in the deepest layers of the epidermis, next to the corium. This process could also be described as the 'out-growing of a hole', surrounded by the adjacent horn of the sole. Blood may accumulate in such a concavity in the sole, but the hole may also be filled with an overgrowth or a protrusion of the corium which is no longer supported. In that case the edges of the horn will press into the quick.

The rest of the claw often reacts by forming excessive horn. The claw grows higher and even more load will be placed on it; the contusion increases and the horny edges will press into the quick even more strongly.

At length the damage will have become so serious and long-lasting that the quick will die off locally in that area.

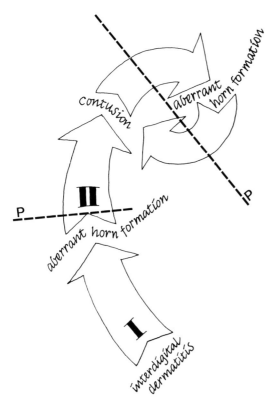

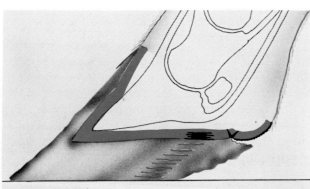

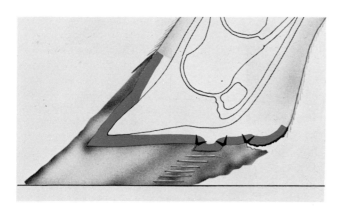

Another vicious circle is created through the contusion of the quick by the hard anterior margin of a deep fissure. This time the horny edge and the *swelling of the bulb area behind* produce a vicious circle: the pressing edge causes swelling and the swelling again increases the pressure (particularly in a claw that has become too high!). At length the damage will have become so serious and long-lasting that the quick will die off locally in that area.

Ulceration of the Sole

Due to various circumstances the quick will in the meantime probably have been exposed to the outside environment. Surrounding bacteria now have a chance to cause a secondary infection of the quick, *where it has*, gradually, *become seriously damaged*. Such a local condition of the corium is known as **ulceration of the sole**. We often find it surrounded by an area of superficially undermined horn.

Contusion and **ulceration** of the sole: stage II in a different phase. As soon as horn formation stagnates an ulceration of the sole can be spoken of, even if there is as yet no secondary infection. In the development of a sole ulcer the contusion of the quick is of primary importance, the contamination being only adventitious.

'Ulcerations' often start as an overgrowth of wound tissue in a lesion of the horny capsule.

Unsuitable horn formation causes contusion with subsequent consequences, usually in *both* outer hind claws at the same time. Treatment as well as prevention of these consequences will have to consist of removing the inefficient horn. This is called trimming or paring (P) the feet. The only way to break vicious circles is with the hoof knife: the sooner the better.

Deviating Postures and Lameness

In the course of the second stage of the heelhorn erosion syndrome contusion and inflammation of the quick cause pain. The unequal distribution of weight-bearing in the hind foot and the heelhorn erosion fissures consistently cause pain in the *outer* hind claws. This is clearly seen in a herd of cattle:

- The animals stand uneasy, occasionally trip, walk sensitively, or lie down too often.
- They adopt weak, cow-hocked or abductional positions; the outer claws remain overloaded (page 33).
- Lameness shows in one or both legs.

There will be a fair number of deviating stances and sensitive animals and a few or several crippled cows, the number of them increasing with age. However, the older animals often have fewer problems – they are the survivors with strong legs, which may never have had heelhorn erosion.

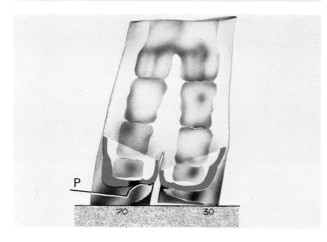

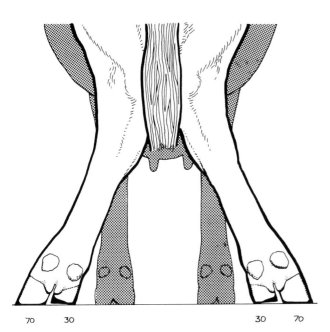

Although it varies widely according to so many circumstances, this picture is an inherent part of intensive dairy farming under the housing and environmental conditions prevailing in so many European countries.

Regular and correct foot care will keep the situation in hand. Trimming may be performed to keep control of the second stage; foot baths can be used to fight the initial stage of the interdigital dermatitis.

If treatment is administered early enough, claw lameness as part of the heelhorn erosion syndrome generally takes a fairly mild course, even if serious changes in the horny shoe do occur. The disease itself is a disorder of the epidermis and, given early treatment, the corium will not yet be badly damaged in the second stage. Contrary to this, claw lameness as a result of laminitis is often less easily cured.

Furthermore, the degree of lameness is not necessarily an indication of the progress of the damage in the quick. Thus the examination of a slight lameness may reveal a process that can no longer be prevented from continuing. On the other hand, a severe lameness may sometimes be easily cured.

Simple Ulceration of the Sole

An ulceration of the sole is a contusion of the corium with interrupted horn formation. If during trimming such a spot comes to the surface, the corium is locally exposed: the sole ulcer is then similar to a bruised wound.

We have also seen that the corium may protrude into the cavity produced when horn stops being formed. After trimming it will protrude through the lesion in the horn.

Sometimes the corium is irritated by the edges of the concavity, and has started to form an overgrowth. This proud flesh (a granuloma) will become exposed by trimming and protrude.

A superficial death of tissue in the ulcer wound is also a possibility. After trimming the corium will be found to be covered by a greyish-yellow layer.

All these cases are classified as superficial or **simple ulcerations of the sole**. The term 'ulceration', indicating loss of tissue in deeper layers, is in fact not really appropriate, but cannot be eliminated from foot-care terminology.

In these cases lameness need not yet be severe. Swelling around the claw is absent or limited in the bulb area.

If the disintegration of tissue penetrates further into the corium, or even into the digital cushion, it is indeed correct to describe the condition as an 'ulceration'. Lameness will grow worse and the bulb of the affected claw will really become swollen and painful.

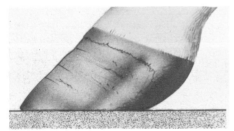

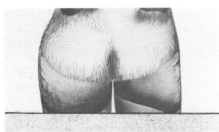

Complicated Ulceration of the Sole

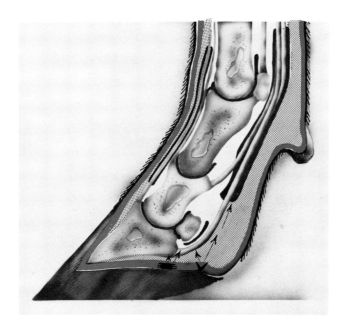

Should the defence and recovery abilities of the tissue be insufficient to stop this process of disintegration in time, the ulceration will become **complicated**. This always involves serious and severe lameness in weight-bearing, accompanied by a clear swelling of the bulb of the claw and, in due course, of the coronet as well.

Initially the painful swelling is found *asymmetrically* in the foot, above the bulb of the affected claw. This may be contrasted to foul-in-the-foot where the swelling is symmetrical and moreover primarily occurs slightly higher, in the concavity of the pastern.

Secondary infections play a part in the development of complications.

The general course of events is easily deduced from the build of the foot.

The first parts to become affected in a more advanced condition are the pedal bone and the deep flexor tendon. The latter, in particular, heals with difficulty. If the necrotising inflammation penetrates the tendon, then the navicular bursa and navicular bone will automatically be affected as well. The pedal joint may still be intact, but very soon it will also become involved in the developing inflammation.

If the necrotising inflammation extends up along the deep flexor tendon, the tendon sheath may become affected too. Before long the infection will reach halfway up the shin bone; lameness is now severe and the whole lower leg is swollen.

An additional problem is that the claw will tilt up if the deep flexor tendon is ruptured by this process of disintegration. A recovery to the normal situation is now impossible.

Indeed, in all these processes, disintegration of tissue will leave scars. This also holds for simple ulcerations.

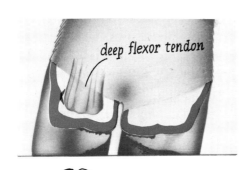

Another serious development in stage II may be the occurrence of an inflammation of the corium on the side of the claw, under the posterior part of the abaxial wall. This may have its origin in a process leading to intracompartmental pressure (pressure within the horny shoe) by tissue swelling, due to a fissure extending as far as this wall. Lameness is evident and the bulb is usually swollen. The centre of this condition, which sometimes shows severe death of tissue, is easily overlooked. Complications, in which the deep flexor tendon and the pedal joint are involved, regularly occur. The condition may be referred to as 'wall ulcer'.

Complications of the 'sole ulcer' phenomenon may be found in various degrees, and it is often difficult to determine exactly how serious and extensive they are. Although their progress to a large extent depends on secondary infections, their occurrence as such depends primarily on the degree of contusion of the quick in the area of the sole ulcer. For contusion strongly reduces the resistance and healing abilities of the tissue.

The longer that unfavourable weight-bearing is able to exert its influence, the greater in general is the chance of complication; the quick in the area of ulceration can no longer stop secondary infections.

Stage II must be treated by trimming *in time*. This trimming, as well as the treatment of the ulceration of the sole, will be dealt with in the next chapter.

Interdigital Overgrowth

One more symptom which often occurs in the course of interdigital dermatitis is the development of an interdigital overgrowth (also called a tyloma).

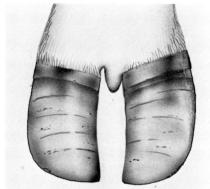

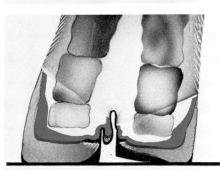

It is well known that chronically inflamed tissue may develop morbid growth. A chronic (lingering) affliction (invasion of bacteria) of the germinal layer may in the end lead to an increased formation of epidermis. Usually the underlying corium will react correspondingly by forming new tissue so that there is more corium and more epidermis in various combinations. The result is an overgrowth between the claws, which is continuously irritated by the claws themselves and the adhering dirt. Irritation in its turn causes more excess growth and again the vicious circle continues.

Major tylomas may become inflamed by injuries and secondary infections and cause lameness. A poultice or disinfecting ointment (in a claw bag) may be effective here, as is the trimming of the axial horn to prevent the overgrowth from being pressed by the claws.

This disagreeable cause of lameness will be prevented by the control of interdigital dermatitis.

Surgical treatment is possible but usually not necessary. This intervention does not come within the scope of this book.

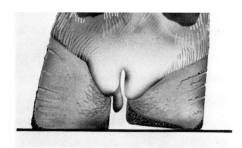

It is remarkable that these tylomas, when occurring in the hind feet, are nearly always situated closer to the outer claw. Sometimes they form an extension of the inflamed and swollen bulb area behind the fissure in this claw.

See also pages 67 and 122.

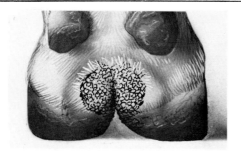

In the region bordering the hairy skin a chronically afflicted interdigital skin may develop wart-like **papilliform overgrowths** (dermatitis verrucosa). Formalin footbaths may be a remedy in simple cases. Disinfecting and desiccative ointments may also be used. In serious cases the best thing is to remove the excess growth by surgery.

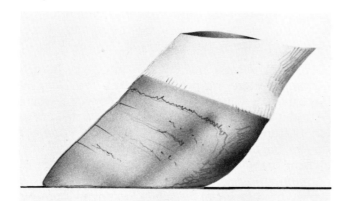

Fore Claws

We ended the previous chapter by discussing why fore claws have fewer problems than hind claws (or rather, than outer hind claws, for inner hind claws also have few problems).

In the tie-up stall there is an evident difference in the occurrence of heelhorn erosion between fore and hind legs. In hind legs it occurs much more frequently and is more severe.

There are more problems in the outer hind claw because of its excessive horn formation.

The inner hind claw, which is normally affected as much by heelhorn erosion as the outer claw, can manage because of the absence of unsuitable horn production: this claw bears less weight and is more evenly loaded.

The fore claws generally show few symptoms of heelhorn erosion, possibly because in the tie-up stall they are kept drier than the hind claws. The fore claws remain healthier!

In cubicle housing heelhorn erosion is widely spread in fore feet too. The familiar bulb malformations occur, but excessive growth of horn is limited. In this respect fore claws may be compared with inner hind claws: stage 1c is mainly absent, and a second stage develops only occasionally. The fore claws remain healthier because of the absence of a systematically unbalanced horn formation.

Only in the case of serious and lasting interdigital dermatitis infections will inner hind claws react by excessive and unsuitable horn production. *This applies to the fore claws as well.*

This problem is increased by the reduction of the grazing period in modern dairy farming. Even then, however, both the inner hind claws and the fore claws are subject to the problems of the second stage only to a limited extent. These problems are mainly reserved for the outer hind claw with its specific unfavourable weight-bearing.

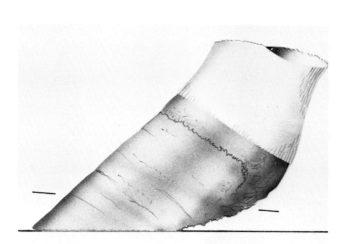

Should any symptoms resembling those of the second stage appear in the fore claws, it is usually a matter of excessive horn formation under the *inner* claw. Here an aggravated typical sole-lesion, or a contusion in a fissure, may give rise to lameness.

Incidence and Control

In present-day dairy farming, heelhorn erosion is very widespread in all sorts of cattle breeds. Fortunately, the course of the disease is usually not as dramatic as would appear from the above description; a great many cases do not reach the more serious stages, and a mild heelhorn erosion appears fairly harmless.

Eventually, however, even a mild course of the disease will, through unsuitable growth of horn, result in an ever-increasing number of sensitive and lame animals, which have difficulty in keeping up in the conditions of modern dairy farming. Even a slight, but continuous overloading of the outer hind claw may cause contusion and, moreover, . . . may promote laminitis.

Not all farms are equally affected, though nowadays there are probably only a few farms in intensive dairy husbandry (with prolonged housing periods) where heelhorn erosion does not occur at all. Differences in hereditary tendency, virulence of the bacteria, housing, management and purchasing policy determine the varying degrees of contamination in farms. Heelhorn erosion is a typical multifactorial disease belonging to intensive dairy farming and with unmistakable consequences for it.

The infection occurs in animals of all ages, but the consequences of the second stage increase with age. The oldest animals, however, often show fewer symptoms: they have withstood the disease.

Bacteroides nodosus belongs to cattle like fleas to a dog. With a certain amount of care and hygiene one may keep the situation under control. Extermination, however, is not feasible for the time being. Heelhorn erosion must have existed for a long time, but 'modern' intensification and neglect of its presence and significance have sometimes allowed the situation to get out of hand.

In the last few years in the Netherlands intensive care has remarkably improved the situation on a lot of farms.

The symptoms have been discussed.

The source of infection is probably the cow.

It has been emphasised that trimming is important to fight the consequences of the second stage; this will be discussed further in the next chapter. In general the results are striking, since the recuperative properties of the corium remain intact for a long time; *Bacteroides nodosus* affects the epidermis only, not the corium. Stage II concerns the contusion of an initially healthy corium.

Measures to fight and control the causative infection will be discussed in the chapters on foot baths and housing.

LAMINITIS

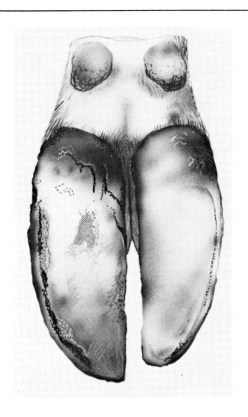

Laminitis (diffuse, aseptic pododermatitis) is a disease which in cattle shows itself mainly by an increasing tenderness, occasionally leading to lameness, in one or both hind legs. Gradually the stance of the legs will become cow-hocked and base-wide. This description of a herd of cattle might also apply to the second stage of a moderate interdigital dermatitis infection. On closer examination of the claws more similarities will come to light, as well as obvious differences.

Chronic Laminitis

Thus, during trimming, we often find, apart from discolorations or sole ulcers in the typical place, *extensive haemorrhages or areas of defective horn* on or just inside the *white line*, predominantly abaxially in the outer hind claw.

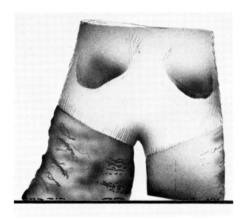

As in heelhorn erosion, the outer hind claw is usually far too high, but it has a *convex horny sole* with, often *diffuse*, discolorations and haemorrhages (beyond the typical place as well).

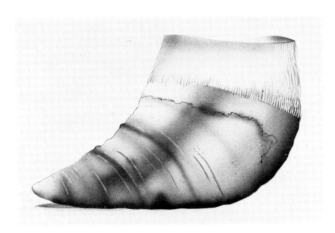

The claw usually has a characteristically *abnormal shape*, which may be described as follows:

— a buckled anterior margin, with a rising toe,
— growth rings that diverge backwards and point downwards,
— a high heel and consequently an insufficiently sloping coronet,
— dull wall horn, and
— a convex sole.

Depending on the degree of the disorder, these deformities occur to various extents and in various combinations mostly in the two outer hind claws but also, albeit less often, in the other claws.

The picture described above, which is frequently observed in Dutch dairy cattle and is well known in a lot of intensive dairy farming throughout the world, is labelled **'chronic laminitis'** ('chronic' refers to its lingering nature).

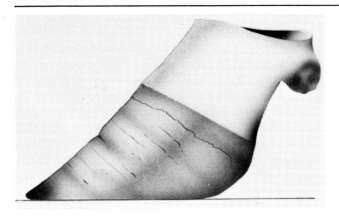

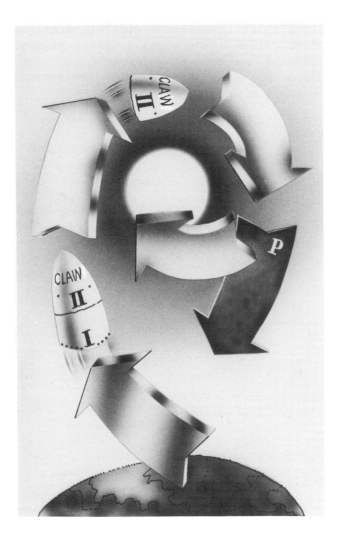

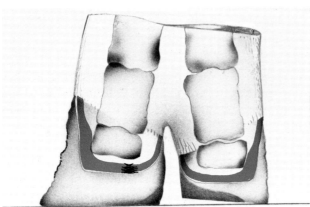

In this disease there are many possible phases, including extremely small changes in shape, of which it is doubtful that these constitute real deformities. However, when these changes in the shape of the horny shoe attract attention, for instance during an investigation of lameness, they invariably turn out to have existed for some weeks or months. It is true that this is also the case in lameness as part of the heelhorn erosion syndrome, but in this case the initial symptoms are known. One of the characteristics of chronic laminitis in cattle is the fact that the initial phases pass unnoticed because they do not attract attention.

Second Stage

Usually the lameness or tenderness are produced by a sequence of events which is **the same as the second stage of the heelhorn erosion syndrome.** The outer hind claws deteriorate as a result of overloading, incorrect loading and secondary infection resulting from abnormal and unsuitable horn formation, which for some reason has started some time ago. Lameness arises through damage to the corium in the **second stage** of laminitis, where contusion (plus secondary infection) and abnormal horn formation keep each other going, *ulcerations* of the corium (sole or wall) being a frequent final symptom.

Only the removal of the abnormally formed horn, which means the suppression of contusion due to pressure and incorrect weight-bearing can break this vicious circle. Trimming is the treatment needed to enable the contused and inflamed quick to heal, and to stop the condition perpetuating itself.

However, laminitic conditions will often be slower to heal than apparently similar conditions in the heelhorn erosion syndrome. The results of treatment are not so sure because laminitis is a *disease of the corium*, which is why the resistance and healing ability of the tissue had already been affected before there was any contusion! Laminitis goes *deeper* than heelhorn erosion.

Unfavourable weight-bearing favours the onset of laminitis, which explains why chronic laminitis is mainly a disease of the outer hind claw.

The diseased corium will produce too much horn. The outer hind claw grows too high and bears excessive weight, especially under the posterior margin of the pedal bone, on the axial side.

This overloading in its turn favours the development of laminitis.

Furthermore, the overloading of a diseased quick results in an aggravated typical sole-lesion which eventually will develop into an ulceration of the sole.

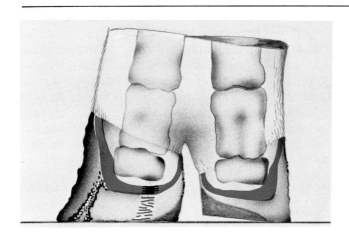

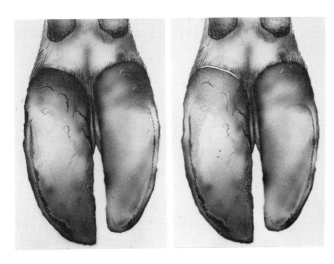

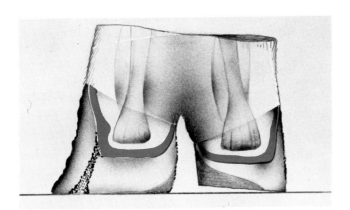

Characteristic of laminitis, however, is the often extensive admixture of blood in the horn near the white line, in the outer hind claw. Through this weak horn secondary infections of the quick in that area will readily break out, particularly as the quick may not be in very good condition due to disease (the laminitis itself) and overloading (outer hind claw). There is possibly a connection with another frequently occurring symptom of chronic laminitis, the defective white line, in which the firm connection between the horn of the sole and the wall is disrupted. The wall will have partly come loose and an inflammation of the corium of the wall may now move upwards and further undermine the horn (*white line disease*).

As a result of these processes in the white line, the horn of the sole may also become undermined. Such an undermining may extend *a long way backwards* under the horn of the bulb.

A deep, easily necrotising inflammation of the corium on the side of the claw, under the posterior part of the abaxial wall, is often the result of an inflammation ascending from the white line and accompanied by swelling and excessive pressure within the horny shoe (compare p. 49, last par.). The condition may be referred to as 'wall ulcer'.

Of course, abnormal stances like a cow-hocked and base-wide posture frequently occur in chronic laminitis.

The complications of necrotising inflammations are basically the same as those described for the heelhorn erosion syndrome.

N.B. Since both diseases occur so frequently, laminitis and heelhorn erosion are often found together. Various symptoms may occur simultaneously, influence one another and cloud the picture.

   This does not make it easier to predict the results of treatment. It is often difficult to determine the seriousness of a condition in the claw before the effects of treatment are monitored.

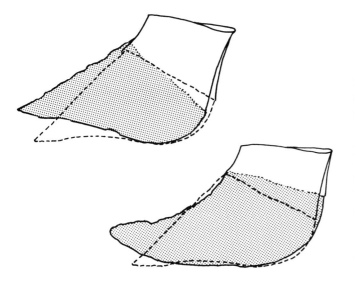

The upward-tilting of the claw which results from damage to the deep flexor tendon, as a sequel to a complicated ulceration, may also be one of the unpleasant complications. This symptom may be confused with the rising toe of a claw deformed by the chronic laminitis itself. In the latter case, however, one can always hold the claws with the anterior margins next to each other when the foot is raised (see page 85). In the case of a tilted claw this cannot be done or only with difficulty; the claws can not be pointed in the same direction. Of course, malformations due to chronic laminitis will not be unusual in tilted claws.

Stage I of Laminitis

The first stage of laminitis is only recognised when we are dealing with an *acute* case. This is a form of laminitis that from the beginning has taken such a violent course that the lameness it causes attracts attention. When this laminitis becomes *chronic* and evolves into the second stage, the symptoms will be those previously described for chronic laminitis, the initial stage of which was not noticed.

For the time being acute laminitis which may become chronic, and chronic laminitis, the initial stage of which was not recognised, will be dealt with as if they were the same disorder. Thus there is only a gradual difference in the course of each: recognition of the first stage depends on how active it is. One may, of course, doubt this supposition.

The clinical picture of laminitis is obtained from observations of acute laminitis that has become chronic.

Laminitis is a **metabolic disorder of the corium and germinal layer**. The condition is characterised by the occurrence of excess *fluid* in the corium of the claw, which may be more accurately described as an excessively high blood circulation (hyperaemia), accompanied by the excretion of blood or fluid from the blood vessels. Oedema develops in the corium; also clots are formed in the blood vessels (thrombosis). Thus, in addition to a generally increased circulation, at the same time there are places where circulation is insufficient (ischaemia).

Abnormal cell formation takes place in the germinal layer and horny substance disappears from the deeper layers of the epidermis: horn formation is upset. In the chronic phase, moreover, fibrous tissue will be formed in the corium reducing elasticity and inhibiting the blood supply.

All these changes strongly affect the functioning of the corium, thereby influencing horn formation in the deeper layers of the epidermis.

Although the moment of origin of each particular case is not recognised, it is assumed that *chronic laminitis* predominantly originates **during the period of parturition**, and particularly **in the outer hind claw**. Later, often during the transfer from housing to pasture, the consequences of the condition are recognised in the claw when a lame cow is treated or when several animals are trimmed.

This onset of laminitis is to be regarded as a *temporary* condition. After some weeks the fluid regime in the corium will recover, to become temporarily disturbed again on another occasion. Sometimes the attack is mild, sometimes serious. Usually scars in the corium and malformations in the horny shoe are left. The scars will eventually lead to malfunctioning.

Older animals will have had more chance to develop laminitis and an increasing number of aberrant gaits and stances will be found among them, particularly as a result of painful outer hind claws. On the other hand the oldest animals may have the best claws because they have shown themselves to be genetically less susceptible to laminitis.

One plausible factor among others is that the oedema of the hind limbs at the time of parturition weakens the corium of the hind claws. The fact that the outer hind claw is particularly subject to unpleasant symptoms can only be explained by its unfavourable weight-bearing.

Indeed, continuously unfavourable weight-bearing on its own is thought to be a cause of laminitis.

Thus the mere housing of dairy cattle promotes the origin of laminitis in the outer hind claw.

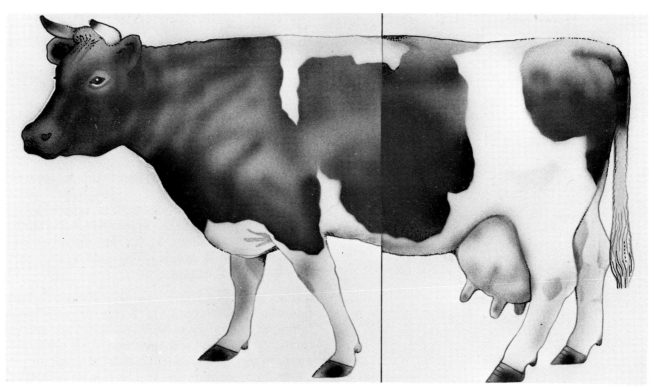

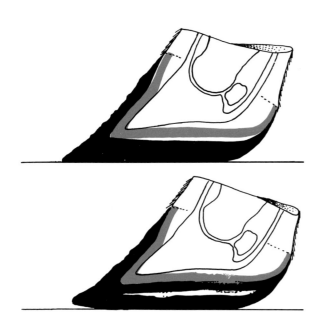

The first things to occur are haemorrhages in the weakened corium and a disturbed horn formation. Visible symptoms (tenderness, lameness) either go unnoticed, are neglected, or are absent. Then, in the course of weeks or months, chronic laminitis develops, with malformations and unsuitable growth of horn, and gradually the condition moves into the second stage. Horn lesions are the result of temporarily interrupted horn formation. Now the corium is *mechanically* damaged or contused, whereas the cause of the original process has long since disappeared. In the second stage of laminitis secondary infections of the weakened corium play a part and ulcerations of the corium are not uncommon.

By regular trimming it is possible to alleviate the eventual changes in the structure of the claw. A considerable part of these changes, however, will prove to be permanent.

Occasionally the course of the first stage in the outer hind claw is so serious that a large-scale excretion of fluid temporarily 'soaks' the epidermis off the corium. Once horn formation is restored this gives rise to the origin of a 'double sole'. This process may sooner or later be accompanied by obvious lameness.

It may also be the case that in the first stage an obstruction of blood vessels, or an extensive haemorrhage, causes the corium to disintegrate locally, thus interfering with the horn formation. In that case a **first-stage sole ulcer** is formed, accompanied by lameness. (See 'subacute laminitis', page 62.)

Multifactorial Disease

The disturbance of the fluid regime in the corium, called laminitis, is often described as forming part of some clinical picture. Then, laminitis is in fact regarded as a *symptom* of various diseases which have in common an upset fluid regime in the corium – just as a rise in temperature is a symptom of various ailments. How the disturbance is brought about is not yet fully understood; there may be several origins.

The concept of laminitis as a symptom partly underlies its *prevention*: 'Keep the cow from falling ill, keep her as healthy as possible, and the threat of laminitis will decrease'. It is difficult to keep highly productive animals in good health, when they are required to digest a lot of concen-

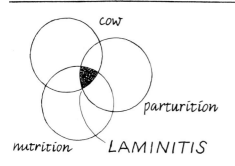

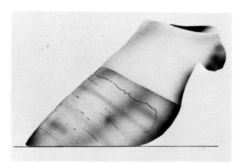

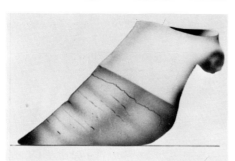

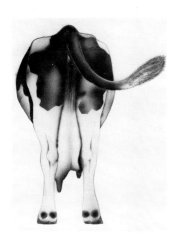

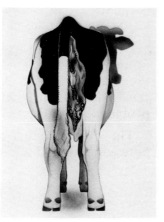

trates and are housed for much of the time. Hereditary tendency – nutrition – housing – parturition; only a slight disturbance of the balance between these various factors may result in illness. Laminitis is not caused by only one thing, it is a multifactorial disease. This subject is also dealt with in more detail in Chapter Five, 'Nutrition'.

The following factors can play a part, usually in combination:

- **Parturition**, which is considered a 'healthy' disease. This annual event is a recurring threat to the outer hind claw through a 'revival' of laminitis. Sometimes the changes in the horny shoe, so characteristic of chronic laminitis, come to light after one parturition only (in heifers, too); sometimes the changes come about quite gradually over the years. The latter looks like an almost 'normal' change in shape during life; a 'normal' consequence of domestication, to which the word disease can hardly be applied. The transition between this and a real laminitic claw is not clear; there are an infinite number of transitional stages between normal and abnormal.

  The clearer the external changes, the more changes in structure are to be expected internally. This implies a weakened claw with an increased chance of contusions.

  Parturition is a debilitating event, in which illness and over-burdening may become decisive factors in favouring laminitis.

- Excessive **udder oedema** indicates a condition of excessive oedema of the hind-quarters. This condition is conducive to the onset of laminitis.

- Violent **bacterial inflammations**, e.g. of the udder, are mentioned in connection with laminitis. This need not necessarily be during parturition, but in that period they will be all the more dangerous.

- Retention of afterbirth (**retained foetal membranes**) is considered a possible cause of laminitis in cows.

59

- **Digestive disorders** are strongly implicated in laminitis, and especially around the time of parturition.

- **Acetonaemia** has been connected with laminitis, as well as

- **Fungi poisoning** through mouldy food.

- Alone, or in combination with other symptoms, laminitis is closely connected with **nutrition**, especially in the period of parturition.

- Lastly, **varying weight-bearing** and **continuous overloading** of the quick may promote laminitis. Thus the specific weight-bearing problem of the outer hind claws and hard floors play a role in originating the disease. Heelhorn erosion makes things worse, as unduly high outer hind claws are prone to laminitis at the time of parturition.

In this situation an effective preventive measure will be to trim the cows when they are being dried off.

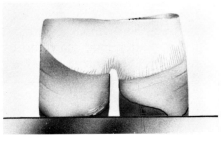

Chronic laminitis may in one season lead to such serious changes that it is correct to speak of illness. On the other hand, as mentioned before, minor laminitis-like changes in shape, particularly those occurring in the outer hind claws and getting worse over the years might be considered 'normal'. At least this applies to highly productive cattle under current housing and feeding conditions.

Parturition and udder development entail a good deal of oedema in the hind-quarters, which could easily lead to changes in the corium of the outer hind claw, tightly confined and exposed to greatly varying conditions of weight-bearing. In-effective feeding and additional diseases may develop this process and eventually lead to a lameness problem classified as laminitis.

Intensive dairy farming demands more milk and less labour, and therefore requires a healthy cow.

Keeping a cow in good health, however, requires optimal care, which means labour. Alternatively a breeding policy can be followed which aims to raise cattle which are not disposed to foot problems – but this may affect milk production.

The answer lies in obtaining the correct balance between care and selective breeding.

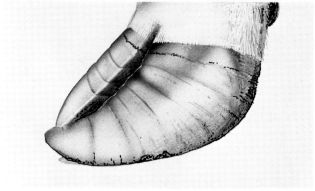

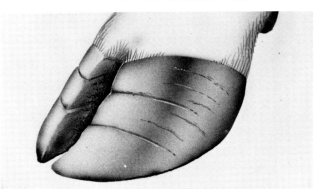

Fore Claws

The hind feet are predisposed to the development of chronic laminitis. Their functional anatomy causes this problem to be predominant in the outer hind claw.

Chronic laminitis largely spares the fore claws which are scarcely affected by parturition. This is one of the reasons for the absence of large-scale problems in the fore claws.

Fore claws remain healthier!

Acute Laminitis

Acute laminitis may occur in fore feet or hind feet, and even in all the claws at the same time. Its onset is sudden and violent, and is always associated with severe lameness. The animals can hardly stand or walk; impact on the claws causes a lot of pain. The symptoms in the corium are the same as described under stage I of laminitis but they take a sudden and violent course.

Stage I of acute laminitis simply cannot be overlooked.

It has to be assumed that stage I of chronic laminitis, the form most common in dairy cattle, takes a less violent course with little or no lameness. It is only in stage II that problems become apparent, caused by local pain from contusion of the corium.

In acute laminitis, on the other hand, the severe first-stage pain originates in vast areas of the corium, and might be connected with disturbances to the fluid regime, which play such a dominant part in this condition.

All kinds of transitional states are possible, ranging from really acute to typically chronic laminitis.

The first stage may vary in length and the second stage may sooner or later create difficulties.

In the Netherlands acute laminitis in dairy cattle is rarely observed. Treatment is beyond the scope of this book.

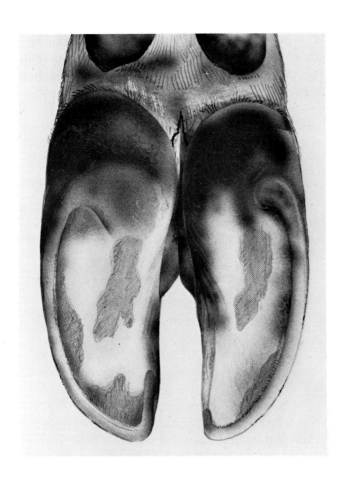

Subacute Laminitis

Regular observation for one or two months after calving may reveal a form of laminitis that we will call subacute: not acute, but not as yet really chronic.

Symptoms are fairly extensive yellow and red discolorations in the horn of the sole and the white line, resulting from a rather seriously disturbed blood supply in the corium at the time of parturition, when blood and tissue fluid entered the horn-producing cell layers of the epidermis. The outer hind claws show the most symptoms, but some symptoms will also be apparent on the other claws.

'Acute' lameness has not been noticed at the time of parturition. Now, several weeks later, lameness may not be too bad, but often a tender gait is noticeable. There are no deformities of the claw yet; later on, the shape of chronic laminitis may arise but not necessarily.

Once the situation has eased, the fluid regime being restored as far as possible, the discolorations grow off with the horn. The outer hind claws in particular will show the 'normal changes in shape due to parturition'. However, really buckled toes can also be expected.

In subacute laminitis ulcerations of the sole regularly occur near the posterior margin of the pedal bone in the outer hind claw. The inner fore claw, too, gets its share especially in the form of haemorrhages in this area.

These sole ulcers cannot be regarded as the direct consequences of overloading, because as yet there is not so much excessive horn growth as there is in the second stage of heelhorn erosion and chronic laminitis. It is more likely that the factors responsible are the combination of 'typical sole-lesion' and poor blood circulation (ischaemia), the latter resulting from the causal disease (stage I) rather than from contusion (stage II).

Subacute laminitis may pose a herd problem, occurring in cattle of all ages; not least in heifers, in which haemorrhages can be very extensive, often undermining the complete sole. If outside injuries can be restricted, the healing of these major defects exceeds one's expectations. The future for these animals does not look too bad although chronic deformities may be the result.

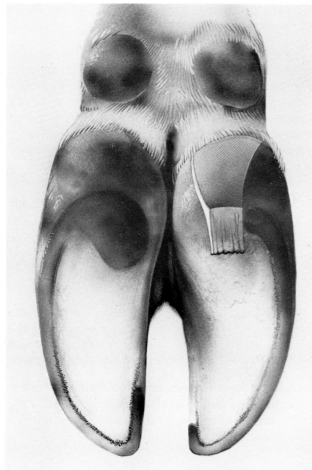

Fracture of the Horny Sole

Fracture of the horny sole means a separation between the horn of the sole and that of the bulb, as shown in the diagram. As the horn of the bulb *joins the sole by a downward slope* (tile-wise) similar to the way that the perioplic horn covers the wall, such a fracture of the horny sole *always* runs upwards and backwards.

The separation *begins* at the surface of the sole and extends backwards to a lesser or greater degree. If the separation reaches the corium of the bulb, lameness will occur due to pressure from loose horn and secondary infection.

The fracture in the sole is usually about 1.5 cm wide. Its depth towards the bulb varies and its width increases towards the back. Eventually the horn of the bulb may become undermined quite widely, the periople of the bulb possibly being loosened over a considerable area.

'Sole fracture' occurs in the first few weeks of the grazing period, in animals *coming from tie-up stalls*. It may occur in all claws and rather frequently. Mostly the fractures are superficial and perfectly harmless and are noticed only during large-scale trimming. The deeper fractures, reaching the quick, may cause severe lameness. However, this is soon healed by timely and effective trimming. (All bulb horn behind the fracture must be thoroughly thinned; loose horn must be removed.)

The outer hind claws in particular and, to a lesser extent, the inner fore claws, may reveal lameness which used to be noticed *especially* in herds *where laminitis was fairly common*.

A general cause could be the *sudden change from dry to moist* conditions, which, as a consequence of differences in water absorption, results in a break in the connection between the horn of the sole and that of the bulb. (This assumption is based on observations of a noticeable swelling of the horn of the bulb occurring during a sudden transition from extremely dry to extremely wet conditions and clearly revealing the demarcation line between the bulber and solear horn.)

Sole fracture is rarely seen in cubicle housing.

If intervention is not made in time and there is a penetrating fracture, lasting pressure by loose horn of the bulb may provoke a serious contusion of the quick. This could lead to ulcers in the sole in the bulb region.

Treatment

Treatment of chronic laminitis must be aimed at terminating the second stage, thus minimising the consequences of the interaction between abnormal horn formation and contusion.

Stage I is past; the cause can only be guessed.

In acute laminitis it is often still possible to trace the cause. An effort must be made to find a treatment that will effectively fight this cause and the pain produced. This, however, is beyond the scope of this book.

The treatment of the second stage of chronic laminitis will be described in the chapter on trimming. Some caution should be expressed about the likely success of trimming this condition compared with the results in the case of heelhorn erosion. In many cases of laminitis the structure of the corium is radically changed, which may result in permanent weakness and lasting abnormal horn formation. *Regular* trimming is often required to reach satisfactory results.

In subacute laminitis there is usually not so much trimming to be done at this stage. In this case good advice on nutrition and housing may still be of use at the eleventh hour.

Incidence and Prevention

Opinions on the incidence of laminitis differ. Prevention is aimed at the possible causes of this multifactorial disease, as discussed in the relevant paragraph of this chapter.

In the Netherlands and many other countries chronic and subacute laminitis occur frequently in all sorts of dairy breeds.

The subacute form occurs equally in all ages, whereas the chronic form increases with the number of parturitions.

Laminitis may occur individually, but often manifests itself as a structural herd disease. It is a 'luxury' disease, connected with high production and intensive feeding.

Laminitis is one of the main causes of lameness and a short production life of cattle in Dutch dairy farming. This does not apply to acute laminitis despite its seriousness in individual cases, because in Holland it is too rare to be considered a real problem in dairy farming.

The *prevention* of laminitis is based on combating those factors that possibly play a part in the origin of this complex disorder. Given certain housing conditions and a certain hereditary tendency, only the **best possible health care** (especially diet and hygiene) will satisfactorily fight laminitis. This includes **periodical trimming**, which should be regarded as a sensible form of prevention, particularly if carried out shortly before parturition.

Trimming will be discussed in the next chapter, while due attention will be paid to health care in the chapters on foot bathing and nutrition.

Subclinical Laminitis

Frequently laminitis shows itself by diffuse haemorrhages, often minimal, in the horn of the sole and the white line, and this is observed during preventive trimming. Because there is no visible lameness this is best described as *subclinical laminitis*.

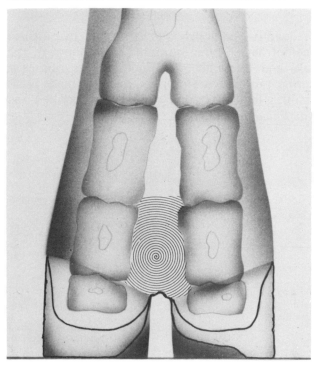

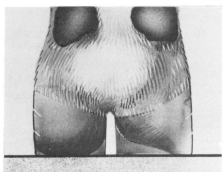

OTHER DISORDERS OF THE FOOT

Interdigital Phlegmon (Foul-in-the-foot)

Interdigital phlegmon is also called interdigital necrobacillosis. (The term 'footrot' is unsuitable because of its confusing usage for different clinical pictures.) It is actually not a disease of the claw but of the subcutaneous tissue *near* the claw, particularly of the tissue between the toes. 'Phlegmon' indicates an inflammation of the supple subcutaneous tissue underneath the tight skin.

Foul-in-the-foot is a sudden and severe inflammation between the toes, presumably caused by an infection from *Fusobacterium necrophorum*. This necro-bacillus (causing necrosis, which means death of tissue) is always present in the environment of cattle and is supposed, under certain conditions, to penetrate the interdigital skin through small injuries and thus to cause the reactive inflammation.

Some people assume that there is a causal relationship between interdigital dermatitis and interdigital phlegmon because these two diseases frequently coincide in cubicle housing. Interdigital dermatitis is assumed to increase the chances of an outbreak of interdigital phlegmon. In tie-up stalls, where interdigital dermatitis infections may take a far more virulent course, with far more damage to the interdigital epidermis in particular, such a connection has never been apparent. The phlegmon is only occasionally found in this type of housing, even where interdigital dermatitis is widespread.

Others regard interdigital phlegmon and interdigital dermatitis as two phases of the same disease: Interdigital phlegmon as the acute phase and interdigital dermatitis as the chronic one; the former occurring more on grass, the latter more in housing.

However, it is doubtful whether this view is correct because the two conditions take such different courses. **Interdigital dermatitis** is a disease of the epidermis of the interdigital skin and the adjacent bulb region of the claw. Stage I of the condition *always* remains *superficial*, even in the most persistent and severe cases. Even though the interdigital skin may be swollen there is never an overall and hot swelling of the foot, or general illness, or a rise in temperature. (The complications of the second stage will not be covered here because they do not constitute a direct consequence of the interdigital dermatitis infection.)

However long the inflammation of the interdigital skin may last, the process will, apart from some reactive swelling in the vicinity, never extend into deeper layers: disintegration is restricted to the epidermis. Likewise a microscopic smear contains no inflammatory cells, which it would if disintegration extended into or through the corium.

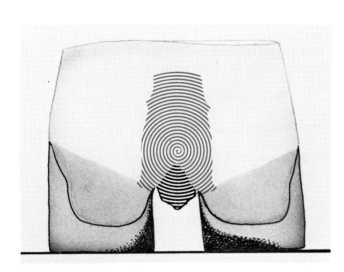

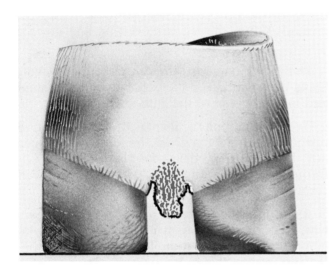

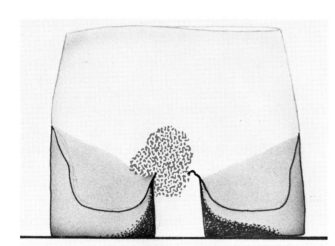

From the onset **interdigital phlegmon** is a *deeply seated* process. Beginning as an inflammation in the supple tissue between the toes, 'above' the interdigital skin, it is accompanied by a hot and painful swelling of the foot, sometimes as far as the fetlock and beyond. The first swelling shows particularly in the middle on the anterior surface of the foot and in the centre of the concavity of the pastern. Further symptoms may be general illness with a decrease of appetite and a rise in temperature.

If the condition does not heal in good time, then the skin and the underlying tissue in the interdigital space will be attacked by the inflammation, which may well be accompanied by necrosis and the formation of pus. This means a deep disintegration of tissue *from inside.* Soon large parts of the interdigital skin are shed, leaving a deep wound which has to be filled with new tissue in order to heal. This often leads to an excessive formation of new tissue, which, after being covered with epidermis, can be seen as an *interdigital granuloma* (also called a tyloma).

Interdigital dermatitis is a superficial bacterial process. Interdigital phlegmon is a deep bacterial process that may burst to the surface.

Interdigital dermatitis has a specific influence on horn formation in the bulb area.

In a severe interdigital phlegmon the horn in the coronet may become locally detached due to the exudation of inflammation fluid. This can be caused by any local inflammation, and is not specific. Interdigital phlegmon in fact does not influence horn formation.

However, where the horn becomes loose the horny rim in the coronet may exert pressure on the underlying tissue, inflamed and swollen as it is. This in turn may increase the swelling, causing the pressure to become greater. . . . A **second stage** begins.

A rim of detached horn, like an 'ingrowing nail', may cause the swelling and inflammation to continue. Thus, even after the actual phlegmon has gone the horny rim may function as the cause of a necrotising inflammation. This situation is serious because the pedal joint is quite near! The only remedy is to remove the horny rim.

Careful examination and treatment of this extremely painful condition is best carried out under local anaesthesia of the foot. This is beyond the scope of the present book.

Distinguishing between the two important bacterial disorders of the foot is not always easy, because:

—heelhorn erosion and foul-in-the-foot may occur simultaneously in one foot;

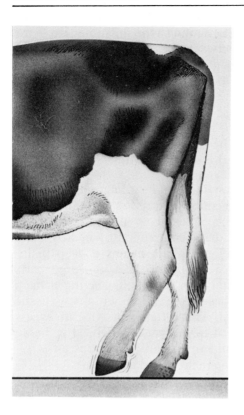

—the inflammation in the second stage of the heelhorn erosion syndrome may be of the same character as the more advanced stages of foul-in-the-foot, namely necropurulent, which means showing disintegration of tissue and formation of pus.

N.B. Digital dermatitis (page 72) may constitute an attendant confusion.

Incidence and Control

Sometimes, at a certain moment, foul-in-the-foot is limited to one or two cases in a herd; at other times we may speak of a genuine outbreak with half the cattle affected within two or three weeks. Usually only one leg is affected, either a fore or a hind leg.

The animals quite suddenly become noticeably lame in the affected limb. A typical symptom is the 'painful' position of the leg: it is held slightly forward, the tip of the claws hardly touching the ground. Often the leg is moved up and down a little.

The foot is swollen, the centre of the swelling being in the middle of the concavity of the pastern and in the middle of the anterior surface.

If the condition is identified in time and if effective treatment is given, foul-in-the-foot need not be a *serious* dairy farming problem. Its frequent occurrence in cubicle housing, however, certainly makes it important.

As foul-in-the-foot is caused by a bacterium which is normally present in the bovine environment and as vaccination has not yielded convincing results, this condition has to be combated by measures which increase **resistance** and promote **hygiene**. As far as possible **housing and yards** should be designed to prevent injuries to the interdigital skin.

Extra *zinc* in the daily ration *is supposed* to have two beneficial effects: increasing the resistance of the skin to the penetration of infection; improving the skin's healing ability if an infection has broken out.

The addition of *slaked lime* to the ground cover in cubicles (about 3 kg of agricultural lime per cubicle) *is said* to have a stabilising effect on the bacterial balance in the house.

Regular formalin footbaths as well as annual or twice-yearly house cleanings are considered effective hygienic measures. The latter is not always easy to carry out.

Stubbly fields and stony paths are thought to increase the incidence of foul-in-the-foot through injuries to the interdigital skin.

Once established, foul-in-the-foot should be treated with antibiotics or chemotherapeutics *as soon as possible.* Inflammation and lameness will then nearly always disappear within one or two days, and disintegration of the interdigital skin will be prevented.

Once tissue disintegration has started, serious complications may arise in the form of extensive inflammations in the foot, originating not in the claw (as in the second stage of heelhorn erosion and laminitis), but in the subcutaneous tissue between the toes.

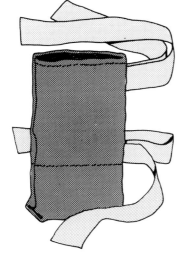

Once foul-in-the-foot has developed, *it must not be treated with formalin*. External disinfection cannot reach the inflammation *in* the foot and, moreover, formalin is totally unsuitable for the treatment of the open wounds which may be formed during the course of a bad healing process. It is much too caustic and could make disintegration of the tissue more likely. The complications of foul-in-the-foot, like those of sole or wall ulcers, must be treated with *mild* disinfectants and rest.

Foul-in-the-foot is discussed in this chapter because it is important to distinguish this clinical picture from the heelhorn erosion syndrome and digital dermatitis.

Stones

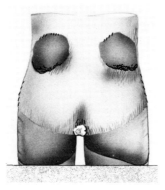

Occasionally a stone may get stuck in the interdigital space. This may cause exactly *the same kind of inflammation* as the interdigital phlegmon. It is easy to recognise the cause of this inflammation (if it is looked for) and of the utmost importance.

After the foreign body has been removed a soda bath (a handful of washing soda in a bucket of water) works miracles. In serious cases, however, after-treatment with antibiotics may be advisable.

Congenital Deformities of the Claws

Inborn deformities of the claws and resulting pain, lameness and abnormal stances form only a small part of the lameness problem in cattle. Although inborn deformities (such as 'corkscrew claw') regularly occur, *numerically* these cannot be compared with the incidence of those deformities *acquired* through interdigital dermatitis and laminitis. There are of course many combinations of inborn deformities and the claw diseases already discussed.

Problems in *healthy*, but abnormally-shaped claws arise if the corium is hard pressed or incorrectly loaded. Pressure by the horn as well as loading through the pedal bone determine the origin of pain, in rest or locomotion, in the same way as during the second stage of the claw diseases. In most cases abnormally-shaped claws will be higher than neighbouring ones and therefore overloaded. A second-stage vicious circle will readily arise, in which contusion through incorrect loading promotes abnormal horn formation. Only *periodical* trimming will keep the ever-returning, undesirable shape of the claw within functional proportions.

Superficial Purulent Inflammation of the Corium

Claw lameness in cattle may also be caused by an *external* injury to the healthy horny shoe touching the quick. These defects, however, play only a minor part in the lameness problem: it is not easy to damage a healthy horny shoe deeply.

A superficial injury to the healthy quick will usually lead to brief lameness, which will spontaneously heal again. The healthy corium has a considerable resistance to secondary infections from the cow's environment.

Should the corium become infected in spite of this, it will usually react by forming pus. This inflammatory product will be enclosed in the horny shoe if the (usually slit-shaped) 'wound' in the horn offers an insufficient outlet. Pus trapped in the horny shoe causes severe pain and lameness!

Although lameness may be severe (as in foul-in-the-foot, but without the swelling of the foot), it is never serious. Discharge of the pus is usually followed by a quick recovery: the corium is only *superficially* damaged. Soon a new layer of horn is formed which may result in a 'double' sole in the area where the pus had accumulated.

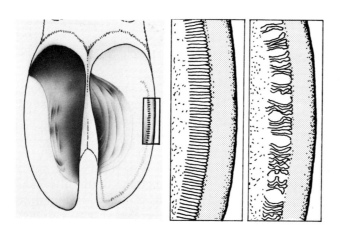

A typical example of the superficial purulent inflammation of the corium (pododermatitis superficialis purulenta) is infection which enters through cracks in the horn as a result of trimming under extremely dry circumstances. This is explained as follows: in trimming, a deeper and softer layer of horn comes to the surface, which becomes drier and harder in accordance with the external conditions. If this process is fast, cracks may be formed. This phenomenon may be particularly noticeable in the horn of the white line, due to drying and deformation of the so-called horny laminae. It is advisable not to trim too thinly under dry conditions.

It goes without saying that a superficial purulent inflammation of the corium will often be found in laminitic claws showing lesions of the horny shoe, especially of the white line. In these cases the purulent inflammation of the corium is as it were superimposed on the underlying laminitic process. In this instance, healing will be unfavourably affected by the disease of the corium.

Serious cases of claw lameness nearly always have their roots in the fact that the corium has already been *thoroughly damaged* before a secondary infection takes place. If in this case it is not the **resistance** (visible by the formation of pus) that dominates, but the **disintegration**, a so-called necropurulent pododermatitis arises. The danger is that, enclosed within the horny shoe, this inflammation of the corium with its disintegration of tissue will spread to the pedal bone, the flexor tendon, the navicular bone, the pedal joint, and, finally the synovial sheath.

The situation is serious if resistance declines and disintegration becomes prominent. However it is not always easy to evaluate the situation.

Sandcracks

A sandcrack is a fracture in the horn of the wall which is in the same direction as the horn 'grows'. If such a crack goes right through the wall and reaches the quick, severe lameness may follow, particularly if a secondary infection causes pus.

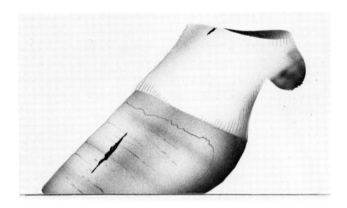

Sandcracks occur more frequently in fore feet. Old cracks rarely cause lameness, except in a very dry summer, when the horn is perhaps less elastic and will more easily crack right through to the quick.

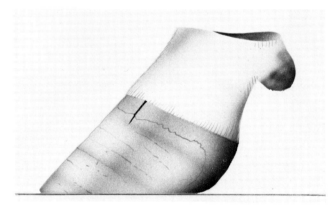

It is assumed that nearly all sandcracks originally start as small fractures in the coronet, growing downwards with the horn of the wall. New cracks will quite often appear in the same place again. The cause is unknown.

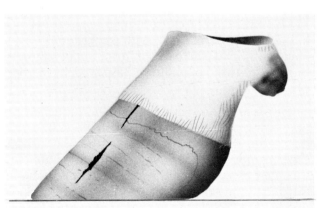

These small cracks in the coronet, though easily overlooked, may cause acute lameness. Immediate treatment with antibiotics should solve this problem by arresting any inflammation of the underlying corium. On the other hand, soda baths can be effective too.

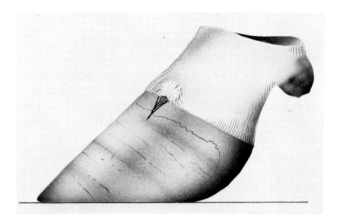

Healing will be delayed considerably once pus has been formed and the corium is damaged. In these cases, the pus must be removed, the corium under pressure must be exposed and the edges of horn must be thinned. It is important to control a second stage in which the horn exerts pressure on an inflamed quick.

In contrast to the situation in a defective sole, the unprotected corium will start to protrude fairly soon, even if the horny edges are well thinned. A dry compressive cotton bandage may be effective in the cases of superficial damage to the corium, as long as this can be prevented from getting wet. (A wet compressing bandage will pinch and is worse than the ailment.) If this is not practicable, a plaster bandage could be used. Topical dressing with mild disinfectants or desiccatives can be applied. In very painful cases treatment should be performed under local anaesthesia.

In practice a less radical treatment may give results: the local removal of a little horn, just enough to release the pus but leaving the corium unexposed; a good deal of shortening and trimming of the whole claw; fixing a block to the healthy claw (in order to remove the load from the affected claw); possibly apply a complementary treatment with soda baths or a clawbag with a poultice.

The latter may work miracles . . .

Serious cases, with necrosis of the corium and underlying tissue, are dangerous because the pedal joint is near. Wet disinfectant bandages and antibiotics are to be used; a block under the healthy claw can be life-saving. Rest is indispensable.

Digital Dermatitis

Under the name 'digital dermatitis' Cheli and Mortellaro (Italy, 1974) have described a disease of the foot in dairy cattle, which is characterised by *circumscribed*, often sharply limited, inflammations of the skin near the coronet. In the Netherlands a similar condition has been observed in recent years.

The typical features are a superficial, strawberry-like, 'open' dermatitis, with superfluous serous excretion, and surrounded by a small epithelial border. More papilliform aspects have been observed.

The lesions are very painful when touched and may cause moderate to severe lameness. In certain herds the disease may spread rather extensively. One or more feet on each animal can be affected, fore feet as well as hind feet. Where it borders the horn the disease seems to accelerate horn production.

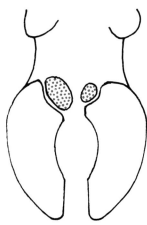

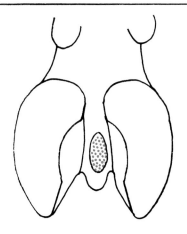

The most likely locations are:

—The bulb region, mostly behind the fissures resulting from interdigital dermatitis; possibly higher up the skin of the pastern.
—The interdigital skin usually on top of an interdigital overgrowth.
—The coronet of the wall, usually near the transition to the interdigital skin. These lesions could be responsible for the onset of defects growing down in the anterior part of the axial wall.

There also seem to be exceptional cases of digital dermatitis as a lesion of the pododerm (the corium) of the claw itself, entailing subsequent local stoppage of horn production.

Up to now the cause is unknown, which means that no preventive measures can be advised. Hygienic housing and formalin footbaths are recommended.

Treatment is possible: a single, intensive topical application (*after cleaning and drying the lesion*) of medicaments containing a tetracycline and **gentian violet** appears to be effective. This should be carried out throughout the herd in all affected limbs in order to control, at least for some months, the spread of the disease.

Unspecific Inflammations of the Skin

In addition to the typical bacterial diseases of the foot (heelhorn erosion, interdigital phlegmon, and, perhaps, digital dermatitis) the skin of the foot is sometimes subject to inflammations which appear to be different and seem to be unspecific. Environmental conditions (humidity, low temperature, chemical components) may affect the skin, leading to secondary infections by environmental bacteria. The horn of the bulb, too, may be unspecifically affected.

Disinfecting, desiccative and hygienic measures usually bring about an early recovery, providing that the *causative conditions are removed*. If the skin is involved treatment should be a mild one; an ointment can be used to treat chaps. *Formalin foot baths will be harmful and should not be used!*

Complications involving disintegration of tissue and formation of pus may occur here as well. Expert advice may be desirable.

73

The Connection between Claw Disorders

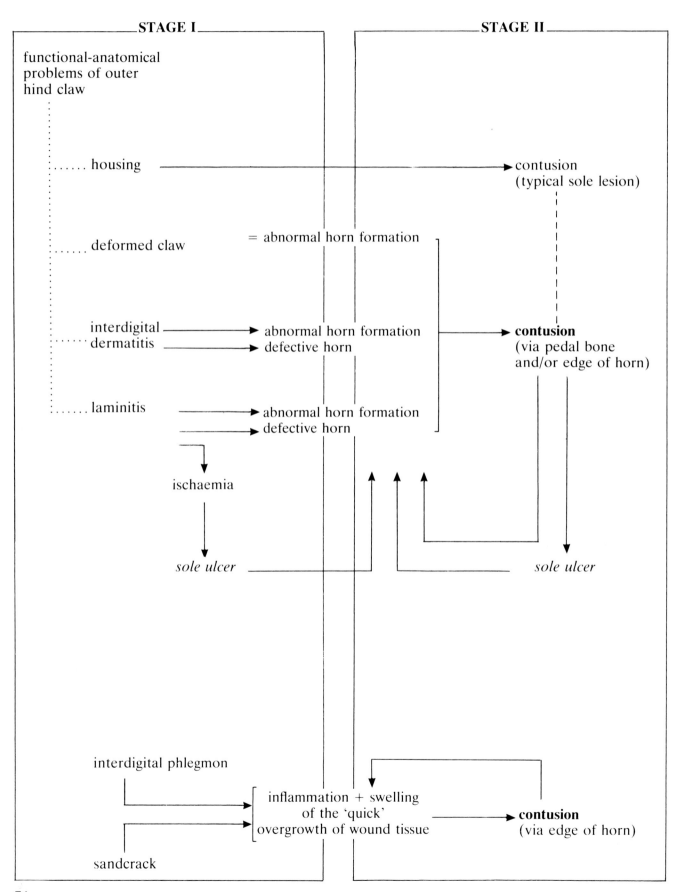

Chapter 3

TRIMMING

AIMS AND POTENTIAL

Trimming here means **to cut and pare the claws in order to make them function as well as possible**. Trimming can also be used to make the claws of cows or bulls neat for showing; however, in foot care the *functioning of the foot* is the main objective. This does not imply that the hooves need not be '**neat**'. 'Normal' implies both neatness and good functioning and it is the normal claw which the chiropodist should keep in mind whenever he does his job.

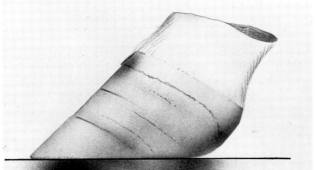

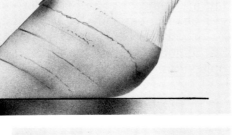

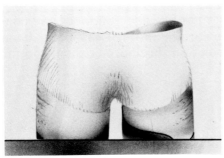

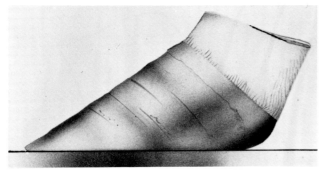

As was explained in Chapter One, 'Structure and Functions', unsatisfactory functioning of a normal claw may result from an excessively long wall, which makes the claw tilt axially and backwards on the hard surface of the house. Thus there is less load on the quick in the weight-bearing border of the wall and more in the axial part of the bulb. The division of weight *within* the claw is less satisfactory.

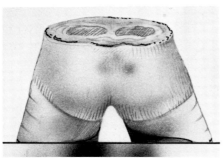

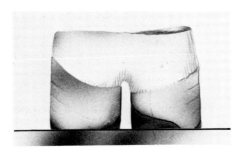

Furthermore, during a prolonged period of housing, the excess work required of the outer hind claw may produce trouble in the healthy foot: one of the causes of the typical sole-lesion is the incorrect distribution of weight *over* the claws.

75

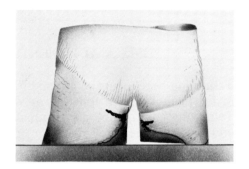

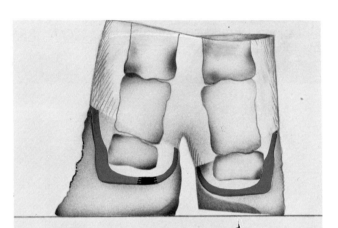

In both cases, which are the results of domestication and housing, it is unlikely that there will be much lameness. Although the animals may be slightly sensitive and may stand and walk less easily, functioning is still reasonably good.

The real functional problems (standing and walking with difficulty and, later, lameness) will arise in *diseased* outer hind claws. *Obvious* overloading causes contusion of the *diseased* quick, in addition to which damage to the corium caused by pressure through edges of defective horn starts to play a significant role. Secondary infections will get a chance, particularly if foreign matter such as dirt or pebbles penetrates into the defective horn and behind a loose wall. To a lesser extent this applies to inborn claw deformities as well.

Poor functioning of claws, mainly in the second stage of claw diseases, is caused by:

- Overloading followed by contusion of the corium through the pedal bone (typical sole-lesion, heelhorn erosion, chronic laminitis).

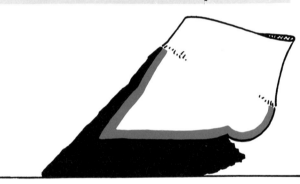

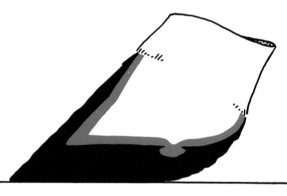

- Contusion of the corium by rims of horn, e.g. rims of fissures in heelhorn erosion and horny edges around sole ulcers.

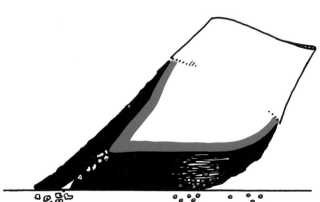

- Penetration of foreign bodies (sand, flints, pebbles) into a defective horny shoe (especially white-line defects in laminitis).

- Infection of the corium caused by penetration of bacteria through a defective horny shoe (typical sole lesions, white-line defects, fissures, or just blood-mixed horn).

The main direct cause of the poor functioning of claws is an incorrect loading of the quick, eventually leading to its contusion.

An incorrect loading can be corrected by trimming:

- by creating a better division of weight *within* a claw;
- by creating a better distribution of weight *over* two adjacent claws;
- by removing and thinning horny rims.

In addition to this, trimming can:

- to a considerable extent prevent the penetration of foreign bodies, particularly in the white line and under a loose wall, and
- create an outlet for discharge.

(Functional problems in the *first* stage of laminitis are less easily influenced by trimming; they are to a considerable extent the result of the causal disease itself.)

Thus trimming comprises the following procedures:

- shortening excessively long walls to achieve a better division of weight *within* the claws;

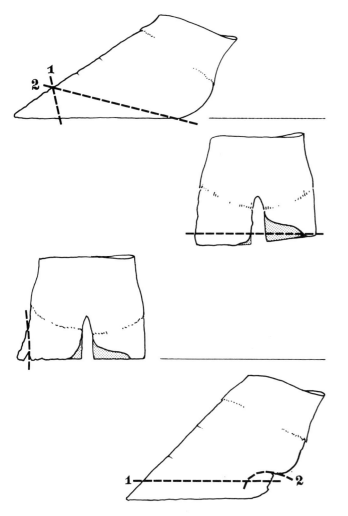

- levelling claws (as far as possible) to achieve a better distribution of weight *over* two adjacent claws;

- removing loose horn to avoid pressure, to prevent penetration of dirt and to release discharge. The release of dead tissue is also made possible;

- thinning edges of horn to relieve pressure.

If treatment is not too late and if the resistance of the quick has not suffered too much, the contused and inflamed quick will normally recover soon. However, it is necessary to stress here that recovery may be seriously delayed if a corium exposed after trimming is regularly damaged from outside: *parlour milking, which involves walking, hampers the healing of diseased claws*!

Thus chiropody does **not** include any treatment of the exposed corium or other damaged tissues. The removal of damaged and disintegrated tissue forms no part of trimming . . . and is basically *undesirable*. The shedding of dead tissue must be left to the body itself; assistance may be given in the form of wet bandages or a poultice.

The 'excision' of sole ulcers may make complications more likely. Only overgrowths of the corium with an obvious narrow base may be cut off after trimming.

Treatment of complicated processes is not discussed here.

Functional Trimming

Chiropody may also be described as **reducing claws to their normal shape and proportions**. This will restore the optimal division of weight. Claws which are too long must be cut, those that are too high must be pared. At the same time, the horny shoe, especially the sole, must be kept strong so that it continues to protect the foot.

Reducing claws to normal proportions in order to restore their normal function could be called **functional trimming**. It involves 'shortening' and 'levelling' the claws. Only *after* this functional trimming may any defects in the horny shoe be treated by removing any remaining loose horn and thinning edges of horn remaining along an exposed corium.

Trimming is always different and yet always the same. Varying quantities of horn will have to be removed from varying places, but always with the same purpose and in accordance with the same rules. The purpose has been explained; the rules will be described in the following pages and summarised at the end of this chapter. These rules have been drawn up for the hind claws. No fixed schedule is produced for the fore claws, which are trimmed if necessary, with the same aims as for the hind claws.

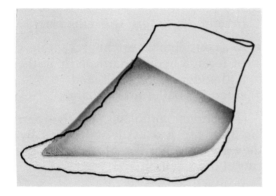

If a reasonable grazing period is observed, fore claws often turn out to control their own length. The 'dead' sole swells up through water absorption and is eventually shed, leaving an unsupported wall of considerable length which breaks off.

Occasionally too much of the wall breaks off, exposing the quick in the tip of the claw. In this case the animal suffers an obvious lameness, but will usually recover spontaneously after a few days to a week. A dangling claw-tip has to be cut off.

Walking to the milking parlour also reduces the length to normal through wear and tear, as happens in most cubicle housing. This in fact renders trimming unnecessary, as the healthier fore claws are also hardly subject to aberrant horn formation.

In some cubicle housing conditions heelhorn erosion may cause considerable abnormal horn formation in the fore claws as well as in the hind claws, especially if the period on grass is reduced. In that case the fore claws will also have to be trimmed.

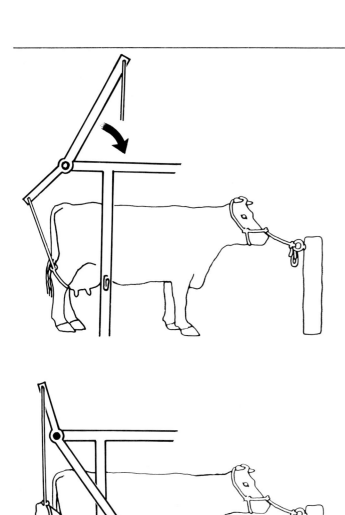

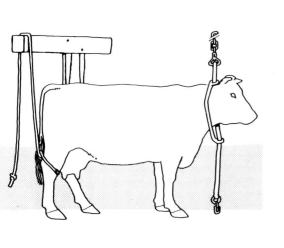

In general, foot care is mainly concerned with maintaining the function of the hind claws. It is often desirable to adjust the stability and distribution of weight *regularly*. Diseases of the outer hind claw may make this an absolute necessity.

Chiropody is a *skill*, just like horse-shoeing, forging, carpentry or pottery, in which practice makes perfect; result and intent have to be compared all the time. Imperfection is inevitable, but should not go unnoticed.

WORKING METHOD

The method of trimming described below is not the only way to deal with a cow and her claws. One technique is strictly adhered to, because adopting **one good** technique is necessary to achieve the skills and results that show that the art of trimming has been mastered.

After some training, certain personal variations may, and no doubt will, be applied.

Fixing the Leg

The hind leg is fixed by raising it nearly perpendicularly by means of a rope tied with a slip-knot just above the hock. In this way the cow retains her four points of support and there is no need to lift anything yourself.

How the leg is raised does not matter, whether it is by a rope across a beam, a pulley or a lever. The important thing is that the noose around the leg should be thick and supple to prevent injuries, and tight enough so as to ensure a good hold. For an effective hold and a convenient working level the hock should be raised up to about a hand's width below the pin bone. If working in a crush, a girth may be placed in the bend of the knee for additional firmness.

The head is best tied short by a halter with a loop that can be pulled loose because, if the animal should fall, the halter rope may be stretched tight; in such a position no cow wants to get up again. Another good hold is provided by fastening the neck with a tight chain or between shackles.

Should the animal fall, there is no reason to panic. It is not supposed to, but it may and occasionally does happen. First quietly lower the

hind leg – mind your own safety. Then untie the head and make the animal rise. In some crushes the cow comes to rest with her neck or throat on a cross-bar or a chain; in that case asphyxia must be avoided!

Disturbance and haste may lead to resistance and falling. Reassuring the cow may prevent a lot of difficulties.

Manual Lifting without Aids

If dealing with a single case, a foreleg may be lifted and treated without aids:

- place your knee against the inside of the leg from the front;
- place your arm along the inside of the leg from behind;
- push the animal with your shoulder onto the other leg;
- push the leg with your knee sideways towards you.

The other leg must, of course, stand vertically on the ground, straight under the cow's body.

However, this unaided method should only be used with quiet cows and is not suitable for bulls, and in the *large-scale* trimming of cows it will prove hard work. In this situation some form of restraint equipment is required. Various kinds of crushes are available for this sort of work.

If the cow is quiet a hind leg can also be dealt with without restraints. This requires a technique similar to that used for the forelegs:

- put your knee against the inside of the leg from the front (your foot will thus stand under the animal, which will not cause any opposition or kicking as long as you press yourself against the cow);
- place your arm along the inside of the shin from behind;
- press the animal with your shoulder onto its other leg, which must stand vertically on the ground;
- turn the hock outwards and pinch the dew claws;
- bend the foot and hold it well below the animal;
- hold your head slightly to the side in view of the hock.

During this procedure your body should be beside the cow's hind leg, not in front of it. Straighten your legs as much as possible and bend the upper part of your body well forward, your elbow supported by your knee. The cow has to lean against an object or should be held by someone to prevent her from moving round.

Unaided trimming is quite feasible as long as the situation is completely under control. A sole ulcer or loose wall can also be treated perfectly well this way, since a good trimming technique will cause little pain: it is horn, not the quick, that is being cut into. If experience and skill are insufficient a good restraint method is preferable.

The manual technique is quite suitable for carrying out a preliminary inspection of the claws and may be useful to make a diagnosis before treatment.

The manual method can be applied only with *co-operation* from the cow. By means of a good technique as well as by talking and pushing, the cow is to be persuaded to shift her weight to one leg, so that the other may be raised. Unrestrained trimming is only feasible if the cow 'understands' what is going to happen.

How to Handle Pincers and Hoof Knife

The following description of trimming will refer to a situation where a rope above the hock has been used to lift and hold the hind leg and the hind claws are being treated.

For technical and safety reasons we work standing beside the hind leg, with our back to the cow's head and our knee or thigh against the front of the lower leg. We are standing not only beside the hind leg, but also against it; as in the unrestrained method. Contact with the cow quietens her and makes working safer, as we are perceptive to her reactions. Any resistance can be felt in advance and one is therefore prepared for it.

This way of trimming also gives the best 'view' of the job, especially when the condition of the claws is being evaluated before and after treatment. Looking down along the claws, the trimmer can see any difference in height between two adjacent claws as well as the position of their bearing surface.

In cutting and paring, the claw is always held in one hand, pincers or hoof knife in the other.

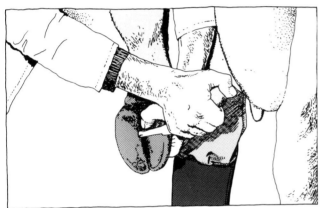

The harder horn of the wall is largely cut off with the pincers and then pared with the knife, together with the softer horn of sole and bulb. The hardness of the horn depends to a considerable extent on the humidity of the ground cover.

Right-handed people put the pincers in place with their right hand (holding them close to the hinge) and squeeze them with their left. Right-handed people nearly always use the right-hand hoof knife. Left-handed people will generally put the pincers in place with their left hand, squeeze them with their right, and then use the left-hand hoof knife. The claw near the trimming hand is invariably cut from the tip upwards, the other one from the bulb downwards. In doing so, right-handed people hold the claw in their left hand and left-handed people in their right, in the following way: the claw that is to be cut in the upward direction is held by the tip, the claw to be cut in the downward direction is held by the bulb. Thus the hand holding the claw is always *on the safe side* of the knife!

An important extra support, adding safety, is provided by the knee or thigh placed against the front of the lower foot.

The technique described above is not the only way to hold and trim a claw. It is, however, recommended and certainly a good way to learn.

The claws are usually not so hard that there is a regular demand for a mechanical form of trimming. There are electric tools available which, however, are dangerous in untrained hands!

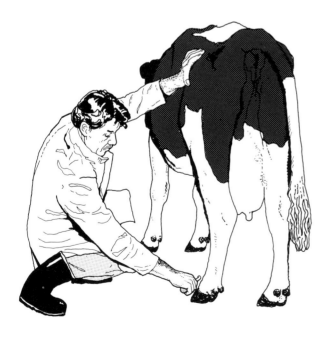

EXAMINATION

'Think before you act.' When applied to trimming this means: **first judge the situation, then start**. The claw cannot be cut or pared until a decision has been made as to what is to be cut or pared.

The examination begins *while the leg is still standing on the ground*, that is before the leg is raised. Dirt and dung must be removed from the claws (put your hand against the upper leg of the cow, move the back of the hoof knife down along the leg and scrape the claw clean).

The things to be examined are the *length* and *shape* of the claws. A difference in height is often noticeable, but cannot be estimated accurately before the leg has been lifted.

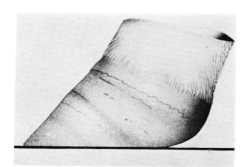

Length

From a technical point of view the length of the claw is the most important starting point. If the toe is of normal length or a little on the short side, it must not be cut any further. If the claws are too long they are to be reduced to normal length with the aid of pincers.

'Normal' length varies with breed and age. A good 7.5 centimetres is a safe size for an adult Friesian cow.

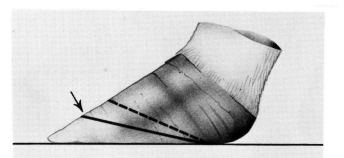

Normal length is best defined as the range between too short and too long. This sounds like an impractical definition; when animals are standing, however, it is fairly easy to decide whether a claw is too long or too short. In practice opinions on this largely agree, placing the normal length at some point in between, with a small margin that does not really influence trimming.

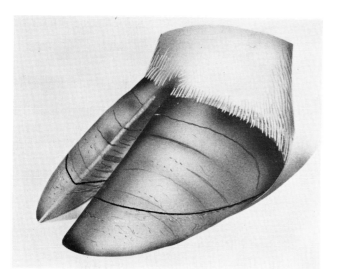

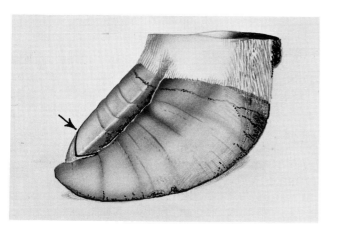

The task of chiropody is to restore claws to their normal proportions. The inner claw may be too long but, apart from this, has usually retained its normal shape; the outer claw, however, often has a more or less abnormal shape which makes it difficult to estimate the correct length. The estimated length of the inner claw, is therefore the best indicator in establishing the normal size of the claws. *Marking this length on the claw*, e.g. with chalk, *will provide the main point of reference for trimming.* The outer claw should be just as long as the inner claw, assuming that the latter has a normal length.

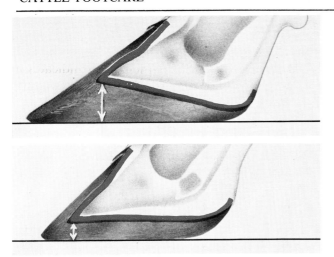

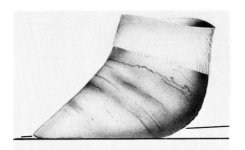

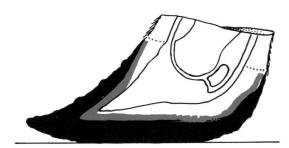

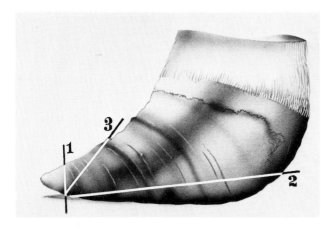

Thickness of the Sole

Length is important because it reveals information about the thickness of the sole, particularly of its anterior part.

In a claw of *normal length* and shape, the thickness of the horn of the sole will range from 5 to 7 mm. Only a small amount of horn may be taken off if desired. It is a mistake to make the sole thin, for the horn is necessary to protect the quick. Where management requires the cow to do more walking, this should be taken in consideration.

A normal length corresponds with a horn thickness of 5–7 mm in the anterior part of the sole. Towards the heel, thickness may increase *with height*.

A *short* claw must not be pared in the anterior part of the sole as there is a minimal horn thickness in that area.

Shape

In the case of an abnormal shape great care needs to be taken because the shape of a claw interferes with our judging ability. The more deformed a claw, the more difficult it will be to retrieve its original shape and proportions.

It is difficult to estimate the *length* of a toe that is buckled after chronic laminitis or has a concave anterior margin due to some inborn abnormal shape, and it is better to play safe.

As a rule, especially in cows suffering from laminitis, the inner claw is the more normal one with the straighter anterior margin, and is therefore a better guide for an estimation of length.

Moreover, a strongly aberrant shape should make us cautious because in such a claw the internal shape may have changed as well. Mostly, however, this change will not be too extensive.

Furthermore we will have to try and recognise the normal shape through the aberrant one, in order to imagine roughly what must be pared away to make the claw as normal as possible. The more aberrant a claw, the more difficult it is to approach its normal form and the more cautious we must be with regard to possible changes in the position of the quick. Some creative imagination is required to broaden the picture of the structure and proportions of the claw; here experience is the best teacher.

Fortunately, in most cases the shape and size of the outer claw can be matched to the more normal inner one.

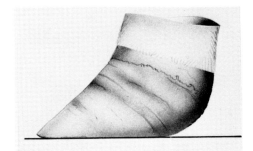

Careful . . . do not touch the quick! Do not make the sole too thin.

The *normal* claw shape is also worth noticing, for, together with a normal length, it indicates a normal thickness of sole. Such a sole needs very little trimming, if any.

Height

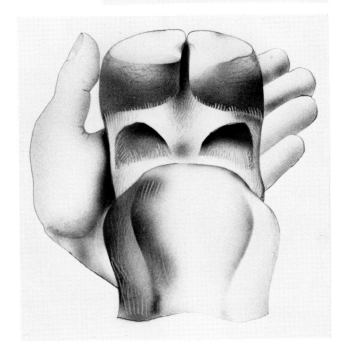

Anyone who knows what a claw looks like will recognise many an excessively high outer hind claw as soon as he looks at a leg in standing position.

After the leg has been raised, the *difference* in height between the outer and inner claws can be noted. This is best observed by looking down over the hock and along the surface of the sole of the claws (see also diagrams on page 86).

For a correct comparison the claws should point roughly in the same direction, which means that the coronets have to be in the same position. This is achieved by placing the anterior margins next to each other in one's hand. If the inner claw is straight and the outer claw concave a compromise must be sought; the rising toe of a buckled anterior

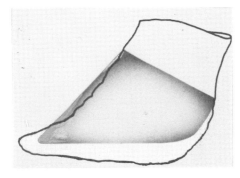

margin must be corrected to some extent before comparing the claws.

It is always necessary to *point the claws in the same direction* if the height or length of the outer and inner claws of a raised leg are to be compared.

In the inner claw the heel is trimmed as little as possible. The aim is to make the outer claw *equally high*, but at the same time keep its sole and heel as strong as possible. This means that the inner claw must remain *as high as possible*.

If the heel of the inner claw is very high (often with deformed horn), its normal height must be estimated and then attained by paring. In practice this comes down to leaving the inner claw as high as possible, provided the *supporting surface is long enough* to the rear. In other words, we must try to preserve as much height as possible and at the same time guarantee sufficient front-to-back stability.

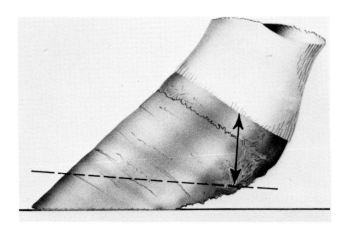

A high inner claw with a short supporting surface is nearly always the result of a serious heelhorn erosion; sometimes of digital dermatitis.

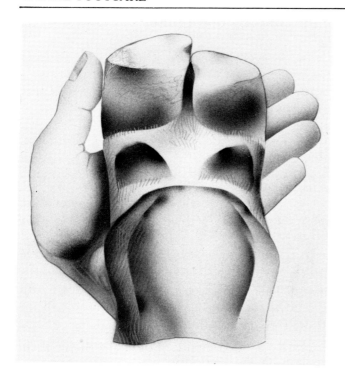

outer claw 'higher'
supporting surfaces 'at right-angles to the shin bone'

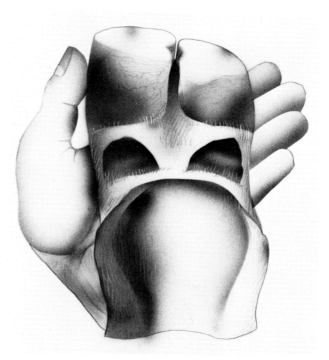

outer claw 'higher'
supporting surfaces 'inclined'

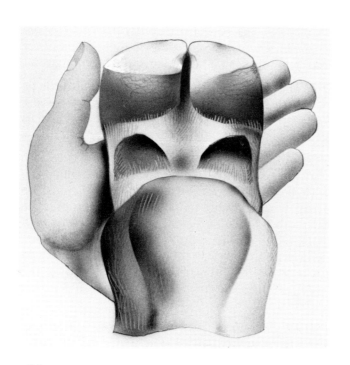

both claws 'equally high'
supporting surfaces 'at right-angles to the shin bone'

WORK PLAN FOR HIND CLAWS

After judging length and height the functional part of trimming follows a fixed pattern:

Cutting the Inner Claw

Cut the excessively long wall of the inner claw as shown in the diagrams: at the tip of the toe, hold the pincers straight (the handles of the tool in line with the bottom surface of the claw); in the abaxial curve, hold them oblique (cutting here will make it easier to pare the claw afterwards); finally, cut flat at the abaxial wall. Towards the heel this abaxial wall must be cut scarcely or not at all, *as the inner claw should remain as high as possible*.

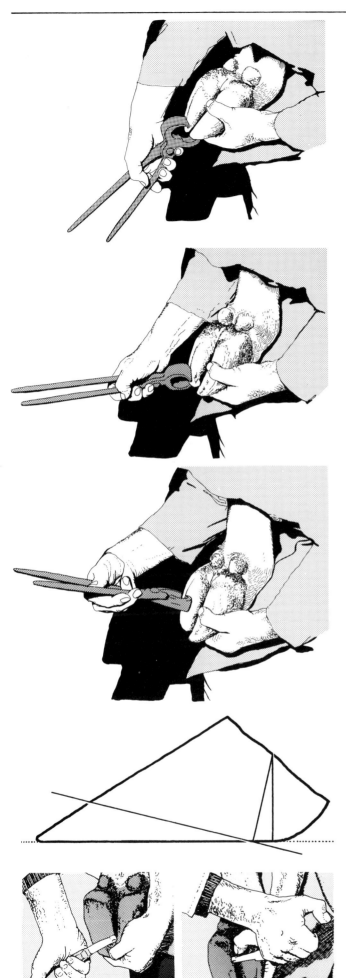

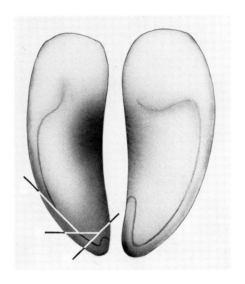

While giving the claw a steeper angle, one may even make the inner claw *somewhat higher* by slightly reducing the length of the supporting surface in the heel.

If the wall is not too long, no cutting should be done.

Paring the Inner Claw

Next the hoof knife is used to pare the bearing surface *flat* (without lumps or indentations) and *in the correct plane*. The hoof knife should really cut and the best technique is to use the arm, without turning the wrist. Paring is safer this way; paring from the wrist leaves an uneven surface and increases the risk of cutting too deep and touching the quick.

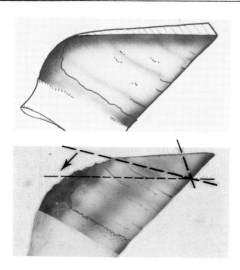

The quantity of solear horn that is to be pared in the anterior part is determined by the margin of the shortened wall. If the wall has been cut rather short, this must be considered when paring the sole.

Heel horn is removed only if this is necessary in order to achieve a fair degree of front-to-back stability. For the rest, leave the claw as high as possible because the purpose of functional trimming is to put more weight on the inner claw by making the claws equal in height. If the inner claw is left sufficiently high the outer claw will not have to be made thin afterwards.

It is not only important to trim flat but also to trim **in the correct plane**, *the new supporting surface being at right-angles to the long axis of the shin-bone in the standing animal.* This guarantees good stability and optimal functioning (see Chapter One, 'Structure and Functions').

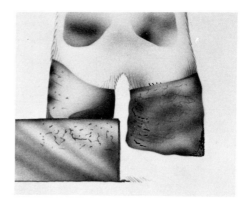

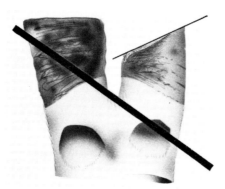

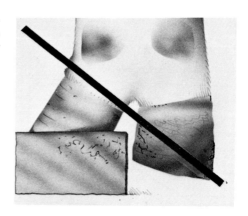

In the raised leg, the position of the bearing surface is best observed by looking down over the hock and pointing the claws in the same direction.

An inner claw that is not too long and has a stable supporting surface is not trimmed. This occurs regularly in cubicle housing.

Height, as well as the position of the supporting surface, determine how well a claw functions. If the inner claw is shortened and pared flat and in the correct plane it will be satisfactory. The supporting area is sufficiently stable also on the flat surface of man-made conditions. However, the weight-bearing ability of the claw is insufficiently utilised as long as the outer claw is higher.

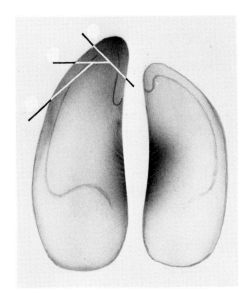

Cutting and Paring the Outer Claw

Cut the outer claw back **to the same length** as the inner one, and cut and pare it back **to the same height**. Then the inner claw will bear its full share of the weight.

Making the outer claw as long as the inner in fact means that the outer claw, too, is cut to normal length. The pincers are handled in the same manner as described for the inner claw. Claws of normal length are not cut.

If the inner claw has accidentally been made too short it is vital to recognise this mistake in time and leave the outer claw a little longer. A claw that is too short is easily pared too thin, especially in the toe area of the sole, where injury to the quick may cause unpleasant lameness.

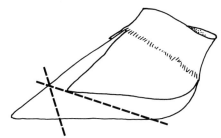

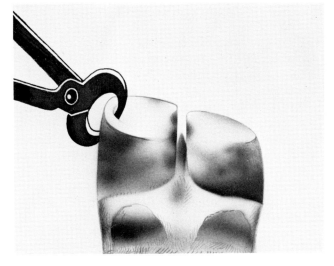

In order to make the outer claw as high as the inner, oblique cutting of the wall by means of the pincers may be continued right to the back. How much of the wall is cut off will have to depend on the observed difference in height between the claws.

Next the bearing surface of the claw may be pared *flat* and in the *correct plane* with a hoof knife, until this surface is at one level with the supporting surface of the inner claw, at right angles to the long axis of the shin bone. Provided the inner claw has been left sufficiently high, this is nearly always possible without making the sole of the outer claw too thin.

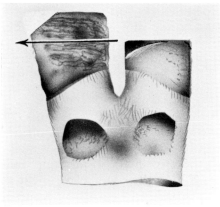

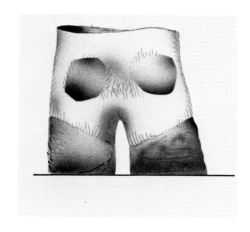

There are exceptions:

- An outer claw of a highly aberrant shape, in which the position of the quick is 'too deep'. This holds particularly for the toe part of the sole in clearly laminitic claws.
- A young and healthy foot, in which the sole of the outer claw is so thin (5 mm), that the usual difference between the outer and inner claw cannot be levelled off by trimming. The sole would then become too thin.
- Very worn claws during the critical period (see page 120); the inner claw has lost too much height.

So, if possible the claws should be made of equal height; but not too thin.

A thin sole may become bruised from the outside, and this irritation may in its turn provoke excessive horn formation.

External conditions should be taken into account. If the cows are either housed or permanently on grass, the thickness of a sole is less important than if they have to use a stony path to reach the milking parlour.

Paring too thin may lead to tender walking as well as injury and secondary infection. (See 'superficial purulent inflammation of the corium', page 70.)

Apart from comparison with the inner claw, there are a few other pointers by which the sole thickness in the outer claw can be judged:

- If, in the posterior half, the elasticity of the sole can be felt by firm thumb pressure, the limit of the protective function has been reached. The sole is then on the thin side because a normal and healthy sole cannot be pressed in like that. The claw should not be trimmed any further, even if equal height has not yet been attained. (This does not count for the curative trimming in the case of horn lesions; page 95.) *In the toe-part of the sole*, elasticity of the horn cannot be used to judge its thickness; here palpable elasticity means that it is already far too thin.
- Beyond the normal thickness, the structure of the horn of the sole occasionally becomes loose and flaky. This so-called 'dead' horn is therefore a good indicator for a safe thickness. It is important to pay attention to this, particularly in the toe area; after the flakes have been pared away perhaps one more millimetre may be cut off.
- A perfectly healthy claw of a normal shape has a sole thickness of about 5 mm. In such cases it is often risky, and possibly unnecessary, to pare the two claws to an equal height because the sole of the outer claw might then become too thin.

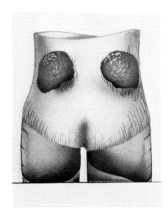

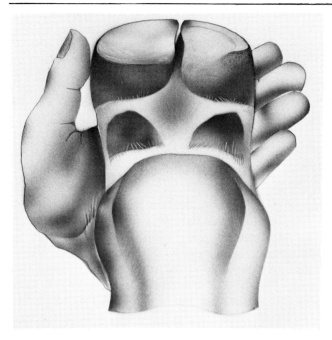

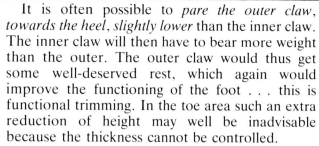

It is often possible to *pare the outer claw*, *towards the heel*, *slightly lower* than the inner claw. The inner claw will then have to bear more weight than the outer. The outer claw would thus get some well-deserved rest, which again would improve the functioning of the foot . . . this is functional trimming. In the toe area such an extra reduction of height may well be inadvisable because the thickness cannot be controlled.

As mentioned above, the outer claw is pared flat and to the correct plane through action from the arm, without turning the wrist. Once again the correct plane is at right-angles to the long axis of the shin bone. The claw then stands upright, so that the weight is shared as evenly as possible: not too much axially, sufficiently to the abaxial side.

Because of its position the outer claw will always tend to become too high.

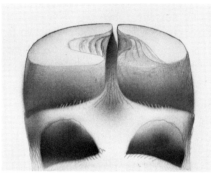

Slope of the Sole

As far as possible the shape of the claws has now been restored to normal; however, the bottom surface of the sole is pared completely flat, particularly in the outer claw. In a normal sole the axial part would gently slope, somewhere in the posterior half of the claw, the slope becoming steeper in the interdigital space.

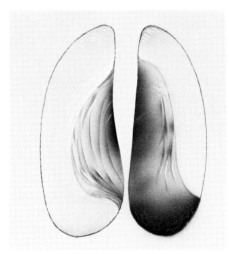

The advantage of this slope is hard to demonstrate. Restoring it, however, guarantees control of possible abnormalities in the area where sole ulcers and fissures regularly occur. Moreover, it will slightly reduce the counter-pressure from the ground in the area of the typical sole-lesion.

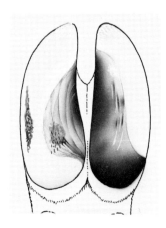

Lesions of the horn that are still present after functional trimming must receive additional treatment.

This does **not** apply to the yellow or red discolorations of the horn where the horny layer is still intact. It is true that these indicate a disorder or contusion of the corium; however, the very aim of functional trimming has been to allow the corium to recover through rest. After functional trimming discolorations of the horn will 'heal' spontaneously, since the cause (contusion of the corium through incorrect loading) will have been removed; damage to the corium caused by laminitis will also heal through rest.

A horny layer must remain, *and be left as strong as possible*, in order to protect the corium against injuries and secondary infections from outside! This certainly applies to the slope in the sole as well; this should not become a cavity.

Discolorations in the horny sole must not be thinned.

LAMENESS

Even without horn lesions and secondary infections, contusion of the quick in an excessively high and deformed outer hind claw may lead to tenderness or lameness. This implies that treatment of a lame cow will not necessarily reveal an open lesion. In the absence of open lesions functional trimming has been the treatment for the lameness.

When trimming a lame cow it is advisable to work on *the lame leg first*. After effective treatment the animal as a rule stands more firmly on this leg, so the second leg may be more easily lifted.

Trimming this good leg is an important part of the treatment of a lame cow. The outer claw of this leg will already be overloaded by the very difference in height that already caused lameness in the other leg. This overloading will increase, because for the time being this leg will have to bear the major part of the body weight (as the cow relieves her lame leg). Functional trimming, therefore, is the first requisite for the better leg to keep it functioning.

If lameness is so severe that the second leg cannot be raised, then housing with a soft surface should be provided, and this leg should be trimmed as soon as possible after some recovery. If the cow is hoisted (so she retains her four points of support) this *undesirable* delay will hardly ever be necessary.

In a case of lameness, extra attention must be paid to checking the claws for defective and loose horn. If such lesions are present, further treatment must be given.

If lameness occurs a few days or a week after trimming or if the existing lameness continues or becomes worse, the claw concerned (*with swollen bulb*) must be examined more closely and if necessary receive further treatment. A serious contusion, for instance, could become infected after trimming, thus causing an undermining which was not yet there at the time of trimming.

Lameness as a result of 'touching the quick' during trimming is not common. Occasionally the quick is touched, but the consequences are slight; nevertheless, this should be avoided. Generally follow-up treatment will not be necessary.

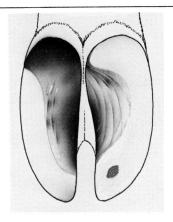

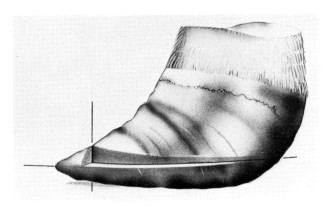

An exception to this rule is a major slip of the knife in the toe area of the sole, which may cause a nasty lameness. A tight dry bandage may help here, as well as a block under the other claw.

Touching the quick during trimming is the result of bad judgement or lack of skill.

FLAT CLAWS

Wide, flat claws with a bearing border that bends outwards, as it were, need a lateral correction towards the size of the normal claw. In making this correction the whole weight-bearing border is often removed, so that the cow comes to rest on her sole. However, this is the only way to make the wall grow again in the normal direction. Whether we succeed depends on how serious the aberration was and how much rest the cow is allowed to take.

If such a lateral deviation is not corrected, the aberration will grow worse and worse, and there is a good chance that a chronic inflammation of the corium of this wall will arise.

The claw shape in question is a result of chronic laminitis.

OVERGROWTH OF THE INTERDIGITAL SKIN

If a major overgrowth of the interdigital skin (tyloma) becomes hard-pressed between the claws a vicious circle (pressure—swelling; swelling—pressure) may readily arise and lead to lameness. This may be corrected or prevented by paring out some extra horn on the axial side of the adjacent claws, in order to create additional space.

It must, however, be done carefully and expertly, for the horn is thin here and injuries are unpleasant in this region.

Take care not to disturb the stable weight-bearing surfaces.

POSTURE OF THE LEGS

Improving the posture of the hind legs, that is restoring their parallel stance, is not the aim of functional trimming. It is, however, to a greater or less extent the result of it, since adjustment of posture becomes unnecessary after the pain in the claws has disappeared.

93

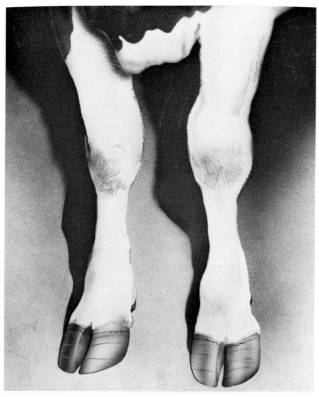

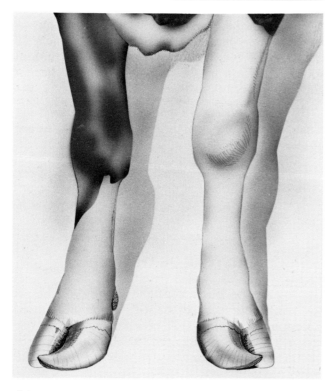

FORE CLAWS

Regular trimming of fore claws is usually not necessary. Even though length and height may not always be ideal, the functioning of the claws is hardly affected. Fore claws are healthier and can stand a good deal.

Where tie-up stalls are used long toes usually break off spontaneously in the grazing period, so that normal length is restored. In cubicle housing wear usually ensures a reasonable size of the fore claws.

In both situations fore claws remain healthier, so there is less ineffective growth of horn, the main reason for trimming the outer hind claws.

Systematic overloading of one claw probably does not occur in forelegs: no anatomical features are known that would indicate a functional disadvantage of one of the fore claws. However, in the inner fore claws typical sole-lesions (discolorations of the horn in the typical place) do regularly occur.

If fore claws do have to be trimmed, the same rules apply as to hind claws: restore a normal length, trim to equal height, flat and in the correct plane. Attention is drawn to the fact that in fore legs it is the inner claw that is usually slightly higher than the outer one, and therefore needs a little more trimming. (The source of lameness is usually in the inner claw.)

The normal length of the outer claw is generally the best starting point.

Often hardly any paring is possible on the sole of the outer claw, because if a claw is healthy, the sole is not thick.

In fore claws the soles are rather flat, so level paring is desirable. The slope of the sole should be prepared in moderation.

In cubicle housing a new phenomenon has been observed, possibly to be regarded as a systematically occurring aberrant growth of horn, owing to these housing conditions. The inner fore claws show a considerable overgrowth, often curving sideways over the front of the outer claw. The condition is accompanied by a 'toe-out' posture of the fore legs (see page 120). It has also been observed that the roughness of the claw bones differs, comparable to the difference in the hind feet, but with the inner claw bone being the rougher.

In obvious cases trimming appears advisable.

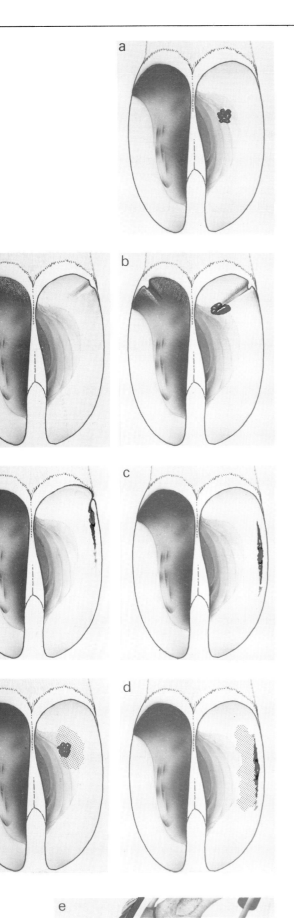

TREATMENT OF HORN LESIONS

After functional trimming, there must be further treatment if the corium is no longer covered by an attached layer of horn or if there is a danger that foreign bodies can penetrate into defects in the horny shoe.

The latter particularly applies to the white line, if wall and sole are no longer firmly connected; it is comparable to dirt accumulating under a loose finger nail.

Horn lesions that may be found after functional trimming can be divided into five categories:

a Sole ulcer lesions where in a certain place in the area of the sole (usually near the posterior margin of the pedal bone) the contact between corium and horn is disrupted due to a locally disturbed horn formation.

In this case the corium usually comes into direct contact with the outside world: an open lesion.

b Fissures due to heelhorn erosion. If heelhorn erosion is only superficial, the corium will not be directly affected; if it extends deeper, with the formation of fissures, the corium will no longer be evenly supported. If a fissure extends as far as the corium, then we are again dealing with an open lesion. This final stage may be compared with a sole ulcer.

c Loose walls, where the firm connection between sole and wall in the white line is disrupted. This usually occurs locally, mainly in the area of the side wall of the outer hind claw. The phenomenon often consists of rather small cracks; occasionally the connection is damaged by a more serious haemorrhage in the horn. Although, so far, support and protection of the quick remain unaffected, penetration of dirt is inevitable.

Sometimes, however, the disintegration of horn goes as far as the quick at the transition from sole into wall. Again, an open lesion is formed, which might be called a **wall ulcer**. The 'wall ulcer' described on pages 49 and 55 includes an undermining of the very posterior part of the abaxial wall.

d More **extensive separations** of horn, where the corium often is only superficially affected. These underrunnings are found as extensions of a sole or wall ulcer, after acute or subacute laminitis, or as a result of an accidental infection (superficial purulent inflammation of the corium). Fairly extensive open lesions may be observed.

e A **sole fracture** allows penetration of dirt between the horn of the sole and the bulb. If the corium is reached, an open lesion is formed, which after some time will be comparable with a sole ulcer.

95

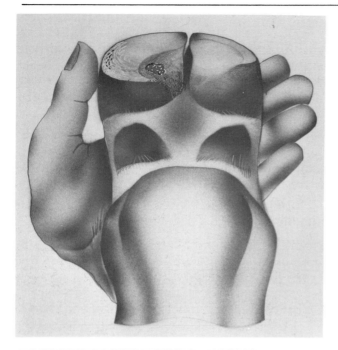

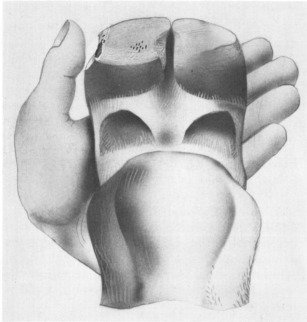

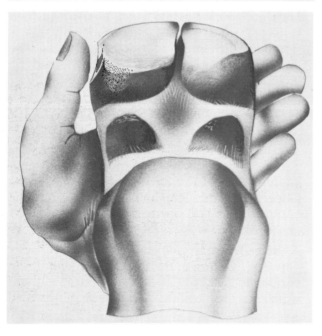

All these processes have in common the fact that the horn itself has become the cause of damage to the quick:

—either because it has become loose,
—or because it no longer supports the corium evenly.

Contusion of the quick through overloading has already been treated by functional trimming. Treatment of any remaining horn lesions, to relieve or prevent additional contusion, should consist of:

• **removing loose horn,** and
• **thinning horny edges.**

Loose horn damages the corium and must therefore be removed. There is often some inflammatory fluid under this horn, which also damages the corium. This discharge can now run off; the quantity of discharge is usually limited.

Furthermore, removing a loose part of the wall will prevent dirt and stones penetrating. A gradual transition to the surrounding healthy part of the wall must be sought. Ideally the wall should be removed as far as the point where the horn is completely attached, but interim solutions are possible, removing a large piece of the loose wall in several sessions.

Even in the case of the slightest defect in the abaxial white line of the outer hind claw the relevant piece of wall should be removed.

A loose wall indicates a damaged corium. Removal of the loose wall will provide extra relief for this part of the claw and give the corium all the rest it needs. The subsequent healing process will result in the formation of a firm new wall.

Sudden transitions in the horn, which might affect the quick like a ridge, should be levelled; hard edges pressing on an exposed corium, e.g. after the removal of loose horn, should be thinned. This is necessary to guarantee the even support of the corium and to prevent horny edges from pressing into the quick. Particularly if the *posterior* part of the wall has come loose and the loose horn has been removed, the edge of the anteriorly adjacent part of the wall must be thoroughly thinned. This is necessary to relieve the pressure of this horny edge on the often swollen bulb tissue underneath.

Pare carefully and thinly in order not to damage the quick! Check by light finger pressure!

Save your sharpest knife for thinning horny rims.

In all these treatments the supporting surfaces of the claws, which have been carefully produced by functional trimming, must be left intact as much as possible! Especially in the sound claw!

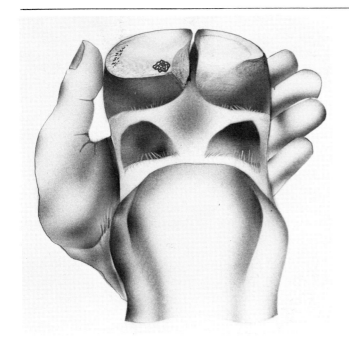

Attention!
As a general introductory treatment of horn lesions it is important to give the damaged claw *extra rest*. So the aim of the preceding functional trimming should not be to make both claws equally high but the healthy claw higher; which means that the defective one should be trimmed somewhat 'lower'.

An additional advantage of this procedure is that loose horn is easier to remove and horny edges are more easily thinned.

This extra reduction in height of a 'diseased' claw is usually feasible, at least towards the posterior part of the claw. In the toe area, however, it is usually only safe to trim down to an equal height, particularly if the claws have been made rather short. In this area, where the corium lies directly on the periosteum, injuries are particularly unpleasant.

In open lesions it is important to reduce height towards the posterior part of the claw as much as possible; although this makes the sole vulnerable from outside. The other claw will take over the weight-bearing so that, in the diseased claw, maximum rest is ensured while the danger of injuries from outside is limited as a result of the reduced load.

External conditions may play a part in this and sometimes a compromise will have to be found. If the corium is not exposed, the sole must not be made too thin. In open lesions there is little choice. It would be ideal to **shorten the wall near a sole ulcer so that it just continues to bear. The minimum pressure by the pedal bone will then prevent the protrusion of the exposed corium**!

Fixing a block under the sound claw constitutes an indispensable additional help ('protection', page 100 and 'blocks', page 102).

Keeping an inner claw high may be incompatible with thinning horny rims. After functional trimming, the inner claw will have to bear more weight. Experience shows that the extra load on the corium through the pedal bone is usually withstood.

However, while keeping a claw as high as possible, it is sometimes difficult at the same time to thin the edges of deep fissures sufficiently. Combined with an increased weight-bearing these may cause contusion and sole ulcers in the inner claw (*swollen bulb*).

This is not very likely, however, and we must give priority to the outer claw!

Nevertheless, this possibility must be considered if fresh lameness occurs some time after trimming. In that case the problem might be solved by *some* further trimming of the inner claw, which will possibly increase the outer claw's weight-bearing. In this case, advantages and disadvantages will have to be weighed against each other.

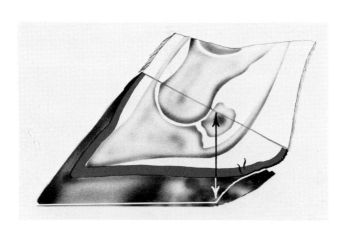

CARE AFTER TRIMMING

Trimming is now finished. The weight has been redistributed within and over the claws in such a way that the healthy quick bears more weight and the diseased quick is allowed some rest. Local pressure caused by horn, discharge and dirt has been removed. Experience shows that this will suffice to bring about recovery.

Deeper damage of the corium, however, may leave scars. Later these can be recognised by discolorations or 'scars' in the newly formed horn. It may be assumed that this will reduce the resistance and adaptability of the claw.

In case of lameness, improvement should be obvious within a few days, provided the resistance and healing ability of the quick have not been affected too much, and if, after the corium has been exposed, **injuries from outside remain restricted**.

After-care of Simple Open Lesions (see page 48)

It thus follows that it is the animal itself which has to accomplish healing of the contused quick. This also holds for simple ulcers, where the animal must shed superficially disintegrated tissue and produce a new epidermis (layer of horn) to cover the wound.

It has been stated already that the wound **must be left alone**: for each fresh injury will further affect the recuperative power of the quick!

Under suitable conditions the animal will be well able to complete its own healing. Hardly any treatment of a simple open lesion of the corium after trimming will be necessary. This also applies to minor injuries of the quick, which cannot always be avoided in trimming.

The effect of using a disinfectant *without* a dressing cannot be great, because the claw is immediately returned to its dirty environment. Furthermore, the effect of a disinfectant *with* a dressing has never been convincingly demonstrated.

Caustic disinfectants impede the healing of wounds! An occasional application of iodine will, of course, not do any harm; formalin (three parts in a hundred) in a claw bag, however, can cause harm in a lesion that is not really superficial.

The necessity of a bandage to prevent protrusion of the exposed corium of the sole is decidedly contradicted by the facts. **After effective trimming, resulting in substantial unloading, simple sole and wall ulcers do not cause the corium to protrude**; any protrusion already present will disappear. If more serious disintegration of tissue gives rise to a general reactive swelling of the quick, protrusion will persevere as the result of an *active inflammation*. This has to be fought first with a wet,

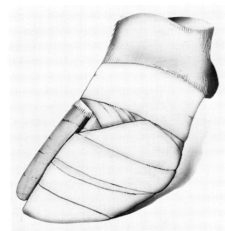

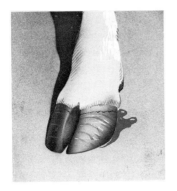

disinfectant dressing, a clawbag with a poultice or a disinfecting ointment, or soda baths. After the dead tissue has been released (the 'purification' of the wound) and the inflammatory reaction has subsided, the swelling will disappear and with it the protrusion of the corium.

With lesions in the anterior part of the sole, where there is no digital cushion to speak of, the corium shows a greater tendency to protrude. Generally these processes are very painful. A pressure bandage to stop this protrusion can hardly be made tight enough to be effective, and it *has to remain dry* to prevent pinching, which is not really practical in farm conditions. Nevertheless, a dry, protective and perhaps looser dressing will certainly be of use in this painful process. *Rest*, however, remains by far the most important factor, though this often presents difficulties in practice. A block (p. 102) might solve the problem. Serious cases should be isolated in the infirmary.

It is doubtful whether 'desiccative' ointments or pastes are suitable for treating a corium that has just begun to heal (to form a new protective layer). The possibility that rapid drying will damage this delicate tissue certainly has to be considered. Once healing is progressing sufficiently, the corium will 'dry up' even without further medical help.

In the case of the disintegration of tissue and inflammatory swelling the first priority is to stimulate the purification of the wound.

In some horn defects, the irritated **wound of the corium** has formed an **overgrowth**. As a final trimming measure, this so-called granulation tissue, *if clearly protruding and with a narrow base*, may be cut off *evenly* with a sharp hoof knife. This is the only exception to the rule that sole ulcers should not be touched. This treatment is not very painful, as granulation tissue contains no nerves. The ensuing haemorrhage, which may seem rather violent, will stop of its own accord. This does not mean, however, that no check need be made.

It is doubtful that this amputation is necessary.

In short, a simple sole ulcer will heal after effective trimming, if necessary with the help of a few soda baths or a poultice to stimulate the shedding of disintegrated tissue. *These aids should be used at intervals during part of the day* (e.g. two or three times a day for half an hour). More serious deformations of the claw, perhaps accompanied by a permanently disturbed horn formation, may have to be trimmed *regularly* in future.

The herdsman should know that in the case of lameness, early and obvious improvement may generally be expected. If this is not the case after a few days or at most a week, the claw must be re-examined!

A diseased claw heals better and certainly faster, if the foot gets some rest. It will heal less readily if conditions are not suitable so that it is regularly exposed to mechanical injury from outside.

Rest without injury is best achieved on grass or in a tie-up stall. Sole ulcers are harder to heal where the cows are centrally milked. Here, the right conditions for healing after treatment are insufficiently fulfilled, as open lesions are continuously damaged because the claw is not rested. This problem is so obvious in contemporary dairy farming that real lame cows in fact cannot be (and are no longer) tolerated. In future claw lameness must be **prevented**.

After trimming, any remaining sole ulcers really must be protected against injury. In most farms, except in a case of emergency, it is no longer possible to fit in a period of necessary rest, which in itself would prevent the injury.

The 'dirty' ground cover, in which the claw has to stand does not noticeably affect the healing of sole ulcers. In both a chemical and bacterial respect, this environment is not demonstrably detrimental to the healing of wounds. *Disinfection is not necessary.* (Minor wounds on the hands sustained in trimming also heal soon without any treatment.) These observations support the view that mechanical damage to the corium is decisive in the origin of sole ulcers, bacterial infections being only secondary.

Protection

Protection against injury from outside may be supplied by dressing the diseased claw, *or placing a block under the healthy one*, or by a combination of these methods. In the first case the lesion is directly protected but the claw continues to bear. With a block, the lesion is protected indirectly as the diseased claw is lifted off the ground. This has the important additional advantage of relieving the claw and thus giving it complete *rest*. Moreover, the cow feels less pain and is therefore able to walk better. *Two obvious advantages over a dressing*!

As is self-evident, completely unloading the outer hind claw also eliminates the continuous *variation* in the weight borne by this claw.

Wrapping up a claw in order to protect it is done either by a dry bandage (keeping it dry is the trick!) or a protective clawbag. A cotton bandage with a thick padding of cotton wool could be used for a dressing, as after a bloody operation on the foot such as surgery on a complicated sole ulcer. Such a dressing, however, is too expensive and impractical for routine treatment of simple open lesions.

A simple protective bandage around the diseased claw, fastened with plaster or tape and with a thin padding, seems more suited for everyday practice. However, the protection it offers is relatively poor and the bandage wears rapidly if the leg is still used.

A dry protective dressing on a 'dirty' sole ulcer, from which some disintegrated tissue remains to be shed, has a drawback in that it inhibits cleansing of the wound by obstructing the discharge. This is generally undesirable, particularly if the dressing remains unchecked for more than 4–6 days.

If in such cases a bandage is applied, a *wet* bandage is more suitable because it not only offers protection, but also *stimulates the purification of the wound*.

In simple open lesions this is not very likely to increase the chance of protrusion of the corium. Such a protrusion is principally connected with the healing phase of the contusion and inflammation of the quick, as well as with the load exerted on the quick. Independently of the bandage, in a still active inflammatory process (swollen bulb), the corium readily protrudes a little, especially if trimming has been insufficient.

Applying a wet dressing is quite possible in practice, but for the sake of convenience the dressing may be replaced by a clawbag with a soda poultice or a mild disinfecting ointment. It is advisable to remove the clawbag every now and then to alleviate the danger of congestion in the foot (if the bag shifts a little, a string may well become too tight).

An easy way to put this on is to press hard against the cow's flank and lift her leg a little forward.

Dressings require checking! After a few days or a week a check should be made to see what is going on under the bandage. Dry bandages should remain dry!

Usually, however, superficial open lesions cleanse spontaneously without any outside help, provided expert trimming promotes healing.

Many veterinarians, and claw trimmers as well, believe that the best way to treat an uncomplicated sole or wall ulcer (the animal is not severely lame, there is no or little swelling in the bulb area above the claw, or in the coronet above the wall) is to leave it uncovered, that is, without any dressing.

The aim of trimming is a sufficient difference in height in favour of the diseased claw (see page 97) as well as sufficient thinning of edges of horn.

A warm soda bath (a handful of washing soda in a bucket of water) for half an hour, two or three times a day will improve a nasty looking condition.

The best after-care by far is a block under the sound claw.

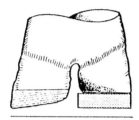

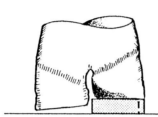

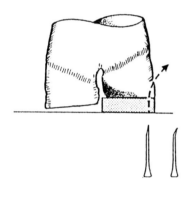

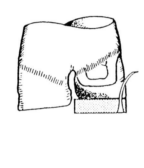

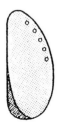

Blocks (see also page 127)

If no sufficient difference in height between the two claws of one foot can be attained by functional trimming, and if the condition and the circumstances require it, a block may be placed *under the healthy* claw. This will provide complete rest for the diseased claw, thereby reducing the chance of injury; at the same time the cow suffers less pain and remains more mobile.

Before attaching a block, the claw must be *pared flat and in the correct plane* while at the same time it is important to keep the sole *as firm as possible*; the former to give the block a stable position under the leg, the latter to achieve a secure hold on a strong bearing surface.

Moreover, after the block has been removed or lost, a high inner claw is still required.

Blocks of hard rubber may be fastened with horse-shoe nails. This technique requires some special skill; putting a little grease on the nails makes them enter more easily. To obtain a solid hold the wall outside the white line must be saved when cutting, for the block has to be fixed into this part of the wall, via the white line.

Another very useful and well-known method is gluing a wooden block to the claw with a commercially available synthetic resin.

A block may be 2–3 cm (about 1 inch) thick. It may be left in place for about four weeks, after which it should be removed. By then it will either be no longer needed or aslant and therefore less useful. Besides, the quick in this claw may become bruised by prolonged overloading. *Increasing lameness following improvement, in the third to fifth week after fixing a block, may be an indication of this pressure contusion in the quick of the bearing claw.* The block must be removed.

Ill-fitting or obliquely placed blocks may cause considerable harm through pressure on the diseased claw. Checking is advisable.

Good-fitting rubber blocks may also have the same effect if the upper axial edge is not rounded in the bulb area.

Instructions after Trimming

Discussing further treatment of an open lesion with the herdsman should be a matter of course for anyone who practises chiropody. The chiropodist has to understand the nature of the condition and therefore is able to appreciate the need for after-care. Depending on what is needed this may comprise rest, wound protection and perhaps the application of a poultice to a 'dirty' wound. If desired, a block may provide the necessary rest and facilitate walking.

Even in the case of an open lesion with noticeable lameness and a swollen bulb, this is the only way to await the result of trimming, at least in the first instance.

If there is no improvement after a few days, a case of lameness must be re-examined.

If trimming can achieve nothing more, expert advice should be sought.

Treatment of Complicated Conditions

A more active approach *may* be adopted after effective trimming, if the condition obviously extends deeper, involving the pedal bone, flexor tendon, navicular bone or pedal joint in the process. These complicated sole or wall ulcers are often accompanied by more severe lameness, swelling of the bulb or even of the whole coronet. The latter is especially indicative of an inflammation of the pedal joint.

These processes must often be exposed further, in order to stop and cure the inflammation with disintegration of tissue (necropurulent inflammation) in deeper layers. This means that a local anaesthetic must be administered, surgery must be carried out, dressings must be applied and, possibly, antibiotics must be given.

Such treatment is beyond the scope of foot care and thus of this book.

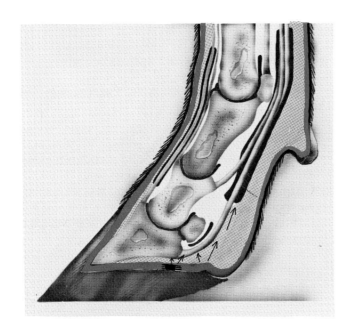

PREVENTIVE TRIMMING

Trimming a severely lame cow, which may have a bad sole ulcer or show a necrotising process under a loose wall, is a difficult task with perhaps dubious results. An operative after-treatment might be necessary; the cow will remain lame for a long time, will have difficulty in keeping up and require a good deal of care, not to mention the loss of production and high bills.

Trimming earlier when lameness has only just started yields better and faster results. Contusion of the corium may not yet have resulted in an open lesion, so that functional trimming on its own will suffice; it will be simple and effective. Early trimming thus *prevents* the more serious stages of a claw disorder.

Trimming at an even earlier phase, which means functional trimming of all relevant animals before they become lame (for instance all milk-producing cows twice a year), turns trimming into **preventive disease control**. *If claw lameness becomes a problem* such a control against claw disorders should be part of modern husbandry, where a lame cow is hard to accept.

Preventive trimming must be done *regularly*, because the claw diseases responsible for ineffective growth of horn cannot be controlled effectively and because prevailing housing conditions promote the unfavourable loading of the outer hind claw.

The ever-arising vicious circles may be broken by regular functional trimming: less overloading—less contusion; less contusion—less unsuitable horn formation; less unsuitable horn formation—less overloading. . . .

Trimming 4–6 weeks before calving seems to be a logical preventive measure against laminitis.

Trimming in autumn is a logical preparation for the housing period.

The advantages are obvious:

—Laminitis caused by over-loading is counteracted.
—Second stages are interrupted in time.
—Contusions and aberrant horn are less frequently formed.
—Lameness occurs less often and gradually there is less horn to be trimmed.
—The herd becomes healthier and causes less trouble.

Whether once or twice a year; whether all milk-producing animals or the older ones only; whether all animals once and the older ones twice, or some animals three or even four times: all this is to be decided by the farmer, possibly after consulting the cattle chiropodist or the vet.

The incidence of claw diseases plays an important role: the more diseases (Stage I) there are, the more aberrant horn formation there will be, and the greater the need for trimming.

On the one side the hereditary tendency of the animals will partly determine the health of the claws. On the other hand this condition can be considerably improved by sanitary measures which control claw diseases. Possible measures against the first stage of the two important claw diseases will be discussed in the next chapters.

SUMMARY OF TREATMENT

Functional Trimming

1. Judge length, shape and height.
2. Cut inner claw to correct length.
3. Pare inner claw flat and in the correct plane, keeping the heel as high as possible while maintaining sufficient front-to-back stability.
4. Cut toe of outer claw to 'same length' as inner claw and, if desirable cut side wall. Points 3 and 4 can be interchanged.
5. Pare outer claw 'equally high', and 'in the correct plane', if possible and desirable, taking into account the circumstances; or slightly 'lower' towards the posterior part of the claw, if this is possible and if necessitated by the condition of the outer claw.
6. Apply 'slope' to sole.
7. Trim other foot.

Treatment of Horn Lesions

7. Pare diseased claw 'lower' towards its posterior part.
8. Remove loose horn of sole, bulb and wall.
9. Thin horn edges and make transitions even; allow space for a granuloma.
10. Block—clawbag—(dressing).
11. *Trim other foot!*
12. Instructions after trimming.

N.B. *Be cautious* in checking for pain!
     Be cautious in checking thinned rims!
     Do not cause extra pain or injury to a tissue which is already damaged.

Planning

Prevention requires planning which should be done during a discussion on the results of a large-scale trimming.

SOLE LESION   ⟶   Consequences

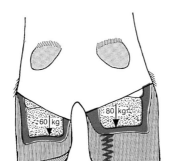

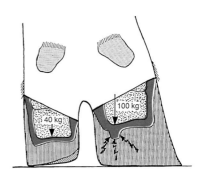

Prevention

Treatment

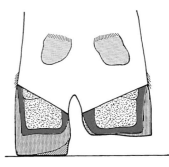

unloaded

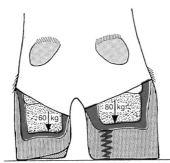

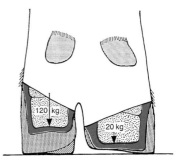

loaded

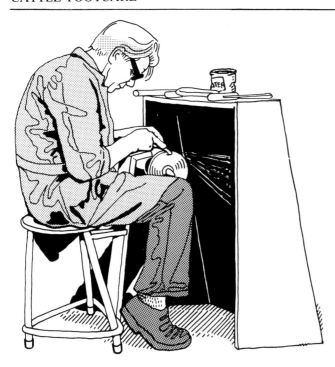

GRINDING THE HOOF KNIFE

A hoof knife should be sharp.

Requisites

- a grinding machine
- a carborundum disk
- a rubber polishing disk
- a fibre brush disk and brushing grease
- a polishing cloth disk and paste.

The grinding machine should be mounted firmly, with a screen behind it to protect against a flying knife. For the same reason the grinder's seat must be protected and should have a support for the hands.

The wearing of safety glasses is recommended.

The machine should be assembled so that, from above, the disks turn away from the operator.

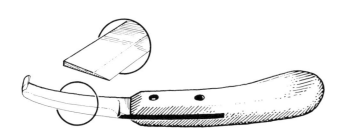

Grinding

When the carborundum stone or the rubber disk are being used the knife must not turn blue. This is very detrimental to the quality of the steel. Press the knife very gently on the disk. Regular cooling in water may be necessary.

New or very blunt knives are first ground on the carborundum stone: the backside *flat*, the concave (inner) side *oblique* (with a slope of 3–5 mm). The hook is made oblique on the outside.

The final touch is given on the rubber disk. Next the burr is removed using the greased fibre brush and the cloth disk is used for a *short* polishing.

A length of rubber or plastic pipe may form a good protective cover for a sharply ground knife.

N.B. Learn the techniques for the safe use of grinders; take no risks . . . trimming is risky enough as it is.

Do not be surprised if your first ground knife cuts badly . . . grinding must be learnt from experience.

Chapter 4

FOOT BATHS

The chapter on claw diseases has demonstrated that in order to guard against claw lameness, aberrant horn formation must be prevented. This means that the diseases responsible for aberrant horn formation must be prevented, which are heelhorn erosion and laminitis.

Abnormal growth of horn under the inner fore claws, a potential result of housing conditions and management, will be discussed further in the chapter on housing.

Measures against heelhorn erosion will be discussed in this chapter, and measures against laminitis in the chapter on nutrition. The effect of housing conditions on the incidence of heelhorn erosion and laminitis will be dealt with in the housing chapter. The chapter on breeding will deal with the question of whether selective breeding may affect the incidence of claw disorders.

Given the cattle and housing on a particular farm, the **fight against heelhorn erosion** has to consist of **regular disinfection** of the cows' feet and their environment: **formalin foot baths** and cleaning the cowshed.

NOTES ON THE CONTROL OF
HEELHORN EROSION

A disease caused by bacteria starts with an invasion of these micro-organisms into a 'host' which has to have a certain degree of susceptibility to these organisms. This susceptibility is determined by hereditary tendency as well as resistance. The animal's resistance again is determined by its living conditions (in the full sense of the word).

- The invasive power of the bacteria (the level of infection) varies from farm to farm; humidity and temperature, for instance, playing a part.

- Hereditary tendency may be different for each breed. Differences in tendency may, however, also occur between farms or between different blood lines and individuals on a farm.

- Resistance depends on housing conditions, food and care, but also, for instance, on gestation, stage of lactation or age.

In each case all these factors combined determine whether the infection will persist or not, but at the same time it is difficult to appreciate the importance of each factor separately.

Theoretically, under given housing conditions the invasive power may be decreased by improving hygiene, and wise management will ensure an optimal condition with the greatest possible resistance. In due course efficient selective breeding might alter the hereditary tendency.

In relation to heelhorn erosion we may note the following:

- Heelhorn erosion is a farm disease. In 1982 there were very few farms in the Netherlands free of heelhorn erosion, but fortunately far from all farms were affected to the same degree. If the incidence on a farm is really serious, the state of health of claws may be dramatically affected.

- There are certain cattle breeds which seem to be relatively less susceptible to the infection, for instance this is said to apply to Jersey cattle. In the Dutch-Friesian breed the disease occurs frequently and is often serious. The Dutch red-and-white breed is also susceptible to heelhorn erosion, although it is difficult to make a comparison with the Dutch-Friesian breed.

- Information about the differences in tendency between blood lines is amply obtainable in practice. Maybe such differences are among the factors that determine the differences between farms. It is also suggested that on a given farm some families and individuals are less susceptible, others more.

  On the other hand if changes in these farm conditions make contamination more likely (for instance a change is made from tie-up to cubicle housing), heelhorn erosion will often spread to nearly all animals.

- The resistance of an animal to the infection is partly determined by housing conditions which, however, not only affect the resistance of the skin of the foot, but also the survival chances of the bacteria outside the host, and thus the chance of contamination. A dry environment is good, since a hard epidermis is more difficult to penetrate while *Bacteroides nodosus* has a small

chance to survive. Warm and humid conditions are bad, since the epidermis softens and the bacterium has a better chance to survive in the ground cover.

Being out on grass has a considerable healing influence on heelhorn erosion.

Some findings have emphasised that the **zinc** content of the food might play an important role in bacterial diseases of the foot: an optimal zinc supply would increase the resistance of the skin. These findings, however, have not been confirmed in Holland until recently.

All in all we may state that Dutch dairy cattle at present are affected with heelhorn erosion. The consequences are clearly visible: aberrant horn formation and lame cows.

A combination of factors, inherent in the development of intensive dairy farming and which probably will not basically change in the near future, have produced this situation.

It is quite probable that in most countries, where the same factors are playing a part, things are exactly the same. Many highly productive cattle breeds are susceptible to this world-wide infection.

Given the current cattle population and farming conditions, the control of heelhorn erosion will, in addition to offering the best possible care, have to consist of attacking the agent of the disease – both within the host and in the contaminated surroundings.

A good measure against such a common infectious disease would be **vaccination**. The antibodies created would help the host resist or even destroy the causal bacterium. To this end these antibodies must be able to reach the agent via the blood stream and body fluid.

However, *Bacteroides nodosus* occurs in the epidermis, where no blood penetrates and only very little body fluid in the deepest layers. Vaccination tests have not (yet) produced results.

For the same reason little good is to be expected from treatment with **injected antibiotics**, as practice has made clear.

Neither vaccination nor parenteral antibiotics are at present considered suitable means of treating or fighting *Bacteroides nodosus* infection in cattle.

A ray of hope could be the fact that, contrary to theoretical expectations, a vaccine against footrot in sheep has proved rather successful. Footrot in sheep is something like heelhorn erosion in cattle, so future developments are awaited. However, the control of heelhorn erosion cannot wait until such developments have yielded results.

As general treatment via the bloodstream has proved to be ineffective, only **local treatment** remains. Theoretically, this too will not be very successful, because the bacteria penetrate far into the epidermis. They are in fact too superficial to be reached through the bloodstream and too deep to be reached from outside. In practice, however, local treatment of heelhorn erosion (with disinfecting and desiccative medicaments) has proved *passable to good*, but the results are usually not lasting. After a temporary improvement or apparent healing, the infection may revive one or several months or even a year later, again leading to aberrant horn formation and lameness.

It is also likely that treated animals will be reinfected by untreated ones.

Given current farming conditions, local treatment is the only possible measure against heelhorn erosion in cows which at the moment has a fair chance of success. This will have to take the form of a **periodically recurring treatment of all animals**, and preferably a **preventive** one; and this means that they must be treated before there are obvious symptoms of aberrant growth of horn. Such regular care is provided simply and at little cost by applying **foot baths** with a **formalin** solution.

It does not appear to be feasible to eradicate heelhorn erosion for the time being. Regular care is needed to control it.

The chances of infection, decreased through periodical foot baths, may be further reduced by cleansing and disinfecting the housing once or twice a year. In fact this is considered to be a good sanitary measure *in general*. Another recommendation is that agricultural chalk is mixed with the bedding.

Treatment of individual animals may be desirable in certain cases, and this may cost a little more time and money. In addition to formalin in a clawbag (three parts in a hundred, **not longer than one hour!**), various other medicaments may be applied locally, but even the most intensive treatment does not usually yield lasting results.

FORMALIN

The correct solution for foot baths is **3 to 5 litres of commercial formalin in 100 litres of water**. (The active component of formalin is 'formaldehyde'. The commercial product contains about 40 per cent of this and is called 'formalin'. Sometimes, however, this is *incorrectly* labelled 'formalin 40%'.)

The advantages of formalin are:

- It is thoroughly disinfecting, even in a bath polluted by dung, urine and earth.
- The bacteria develop no resistance.

- If used judiciously, formalin will not cause pollution of the environment, because, so long as it is sufficiently diluted after emptying, it becomes inactive. What is more, the substance is completely dissolved into carbon dioxide and water.

Copper sulphate may also be administered as a disinfectant in foot baths. In the Netherlands it is thought undesirable because it could affect milk quality; furthermore in this densely cow-populated country it is unacceptable because it contributes to environmental pollution.

Warning: formalin is anything but a harmless product.

- Local vegetation will be destroyed if a lot of used solution is emptied from the bath in one place.

- Formalin is a highly caustic liquid. Eyes should be safeguarded and if they are affected they must be immediately rinsed with water.

- The solution must be correct – splashes of too strong a solution may harm the skin of the cow's foot and its teats.

- Formalin, including the dilution used for baths, is extremely *toxic* when taken by animals or humans! Don't let the animals drink from the bath. Keep children away from baths and keep formalin out of their reach.

- Formalin is not suitable for the treatment of open lesions.

We do not know to what extent the process of bacterial conversion in manure may be affected by regular use of formalin foot baths in cubicle housing. So far as is known there will be no noticeable effects if the baths are not used too frequently.

In short, formalin must be administered *moderately* and *judiciously*.

USING FOOT BATHS

The use of a foot bath can be compared with measures against mange or lice: it is a regularly recurring hygienic procedure to cure, or better still, prevent, clinical symptoms. Extermination is virtually impossible.

The disinfectant must be able to make **good contact** with the interdigital skin. Foot baths should therefore be given *during the grazing* season, because earth does not stick to the claws as much as a mass of caked dung. Advance *trimming* increases the accessibility of a deformed bulb region.

Re-infection from a bacteria-infested environment must be prevented as far as possible. Therefore foot baths must be applied during the *grazing season* (preferably combined with a change of pasture), or when the cows are going into *clean housing*.

It will suffice to restrict bathing to the period just before housing. This will give the cows their best chance of getting through the housing period – so dangerous for their claws – with only a moderate use of formalin.

Bathing in the cowshed is in fact too late for real prevention; cows which are in cubicle housing, however, need this as an *extra*.

In theory, *a grazing period followed by clean housing* provides a good basis for fighting heelhorn erosion. Sometimes this combination is highly desirable for practical reasons as well!

Good results are occasionally obtained in theoretically unfavourable conditions; sometimes a multifactorial disease is difficult to fathom.

There is a widespread notion that bathing is only applicable to the cubicle housing system and that it must take place while cows are housed. This idea is *incorrect*! In cubicle housing the cows can easily and effectively be footbathed in summer, and fighting heelhorn erosion is equally important if tie-up stalls are used.

Stand-in Foot Baths

If the cows are not milked in a fixed place during the period out at grass, a stand-in foot bath must be used in the autumn before the animals return to the cowshed. This in fact refers to the situation in a tie-up stall, where milking is done in the pasture.

In such a foot bath the animals stand in the formalin solution for 30–60 minutes, in order to disinfect thoroughly the interdigital skin and the potentially affected bulb. This may be carried out once or twice; for instance once towards the end of the grazing period (preferably followed by a change to clean grass) and, in any case, once when the cows are brought in to clean winter housing.

A meadow can be considered 'clean' if it has not been used by cattle or sheep for a period of three to four weeks. A cowshed is 'clean' if it has been thoroughly cleaned in the spring and then remained empty over the summer. If any animals have been kept in the house since the spring cleaning it must be cleaned and disinfected.

The dimensions of a stand-in foot bath may be as follows: 5 m long, 1.8 m wide and 15 cm deep. The formalin solution must be 8–10 cm deep, so that the claws are completely submerged. In such a bath eight adult cows can be bathed simultaneously.

Do not forget the young animals!

When administering a stand-in foot bath, remember the following points:

- To prevent any trouble from the formalin fumes the bath should be sited in the open.

- When the formalin bath is emptied, plant life and storage of food must be taken into account. When installing the bath drainage should be arranged.

- Tie the animals so that their heads jut beyond the bath to keep them clear of the irritating formalin fumes; tying will also prevent them drinking from the bath.

- It is advisable to grease the teats after the last milking before bathing.

- If the water is very cold, for instance in autumn when young or dry cows are housed, some warm water should be added for the animals' comfort and to make the formalin more effective.

- If housing is spread over a period, it is not necessary to make a new solution every time; a bath can be used several times over a few weeks. Formalin does evaporate, but only very slowly. However: if the bath is filled with rain it must be renewed; in sunny weather when there is rapid

evaporation of the water the solution might become too strong and the bath should be renewed so that the correct strength of solution is ensured.

Together with bathing, skilful trimming in autumn can be a first-rate form of care.

If during the housing period it appears that the preventive foot bath has been insufficient, the same solution may be used to spray the feet once or twice a week. It is recommended that the teats are greased after milking and that fresh cover is placed on the house floor after spraying.

N.B. Once lameness occurs, formalin must no longer be used! Any open lesions should be cured first by thorough trimming.

A stand-in bath may also serve as a supplement to a walk-through bath; or for the young animals, if the walk-through bath is used for the milking cows only.

Walk-through Foot Baths

If a fixed place is used for milking during the grazing season a walk-through foot bath may be used during the six to eight weeks immediately before the cows are turned into a cleansed house. This is more convenient than a stand-in bath and, if necessary, can be put to immediate use at other times as well.

The results are as good as those for the stand-in bath, as long as it is applied judiciously. This means that after they have passed through the bath the animals should be kept in a dry place for about half an hour. This is the time when the formalin is supposed to be present on the skin and act.

Moreover, the bath should be placed behind the milking parlour, so that formalin fumes do not gather in it, especially if it is closed and warm; it is unhealthy to work regularly in formalin fumes.

If the milking parlour is in the open, the bath can also be situated in front of the collecting area.

A walk-through bath is used twice a day over a period of two to five days. After five days the bath will have become extremely dirty and sufficient treatment will have been given for the present. If feasible the cows should be given clean pasture after the first passage.

A good prevention against heelhorn erosion is three of these bathing periods before housing, at intervals of two or three weeks. This programme may be extended as the need arises, with perhaps a few periods at the beginning of grazing.

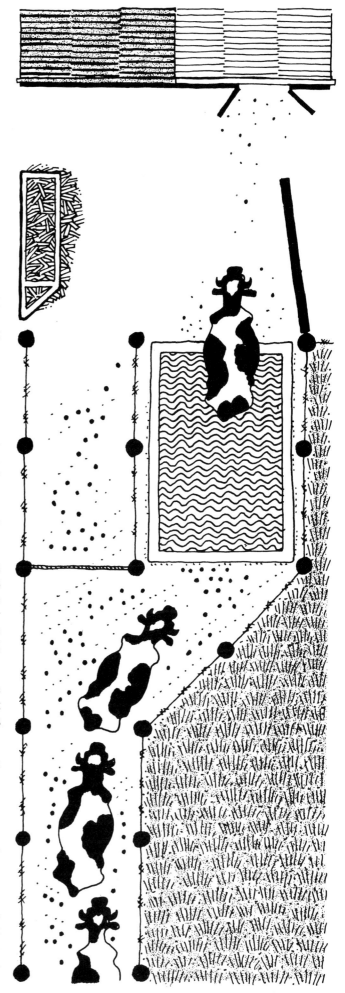

If the housing also serves as a collecting area before milking during the grazing season, it will not be easy to arrange clean housing in the autumn.

Once again it should be emphasised that claw trimming before housing can be a valuable foot care supplementary to bathing.

During a bathing period the animals may indicate by kicking their legs that the skin on their feet is irritated, if so bathing must be stopped at once, otherwise the skin will become damaged. Dry weather with a cold and bleak wind, or frost on the grass, and, of course, too high a concentration of formalin make this condition more likely. After a fortnight bathing may be resumed. (See also page 73, 'unspecific inflammations of the skin'.)

Suitable dimensions for a walk-through bath are about 1 m wide, 3 to 5 m long and 15 cm deep. Again the formalin solution should be about 10 cm deep. It is advisable to put some dung into the bath right from the start, in order to prevent the cows drinking. It rarely occurs, *but a few large mouthfuls may be lethal.* There is no known antidote or treatment.

Because there is a greater chance of contamination in *cubicle housing* it is recommended that *disinfection is kept up* while the cows are housed. A walk-through bath inside the cowshed may be filled with formalin solution once or twice a month. It should be sited behind the milking parlour (*never inside it*), and must not be so short that it rapidly becomes filthy and diluted and loses its effect.

If such an indoor bath is used and the cows stay inside, irritation of the skin of the foot hardly ever occurs.

Of course, the same bath can be used summer and winter.

N.B. Walk-through baths are used for the milking cows. However, it must be ensured that the younger animals and dry cows also walk through it a couple of times; alternatively they must be given a stand-in bath. *If this is omitted there is a chance that control will be less effective because young cows will still be a source of contamination.*

If **no full grazing period** is observed and the housing is no longer cleaned because it is never completely unused, it becomes difficult to fight heelhorn erosion.

During the grazing season the cowshed may be used in various ways: for milking; as a collecting area; as accommodation for the night; or fully throughout the summer.

In all these cases there is insufficient hygiene combined with too limited a use of grazing and its healing effect. This leaves basically three ways to control heelhorn erosion:

- To observe a grazing time, however short, giving foot baths and cleaning the cowshed in the meantime.

- To clean the cowshed while the cows have a stand-in footbath.

- To use the walk-through bath in the cowshed more often. There are certain types of housing where the animals are always indoors, and where such a bath is in continuous use.

The choice in any given situation will be determined by requirements and facilities. A general rule cannot be given.

Some more remarks:

- Most cubicle housing is used as a collecting area in summer. It is recommended that the cubicles are closed and cleaned. The passage-ways can be sprayed clean during the bathing.

- Access to the milking parlour for the cows during the grazing period should be independent of the housing.

- In the construction of cubicle housing cleaning facilities should be considered. These facilities will have to be developed judiciously and with an eye to the cow's welfare.

Finally:

- Small walk-through baths become inactive very quickly because they are diluted by pollution. There is a relationship between the number of cows, dimensions of the bath and duration of activity of the solution.

- Formalin not only kills bacteria; its effectiveness may also be explained by its desiccative effect on the epidermis, which would thus become more resistant to any weakening influence of the environment.

- Formalin becomes less effective if the bath solution falls below 15 degrees centigrade. The implication for a stand-in bath is that it may be desirable to add hot water. If needed, the stronger solution (5 litres of formalin to 100 litres of water) will be more effective.

  In a walk-through bath the temperature of the water is less important: the water adhering to the foot (between the hairs) will always be warmed by the body heat. Here, too, the concentration of the solution may be increased to 5:100 if a better effect is sought.

- A water bath installed in front of the formalin bath in order to wash the feet before bathing might dilute the formalin bath and will certainly dilute the formalin solution adhering to the cows' feet. This could considerably reduce the effect of the foot bath.

- To deal with a problem of heelhorn erosion on a farm it may be advisable to remove the most seriously affected animals from the herd. These animals do not heal easily and might keep the contamination going.

- In addition to treating and fighting heelhorn erosion with formalin baths, regular and skilful trimming is of the utmost importance to keep the claws of dairy cattle healthy. They should be trimmed every year before housing and certainly must be in winter or spring if a lameness problem arises. *Then foot baths must wait* until sole ulcers or other lesions have been treated and are healed.

- Do not treat an interdigital phlegmon with formalin.

- It seems wise to check and bathe newly bought animals before allowing them to join the rest of the herd.

- Footrot in sheep is related to interdigital dermatitis, so it is also advisable to put your sheep through a foot bath.

- On a national level I believe that disinfecting foot baths should be placed at the entrances to shows, cattle markets and similar institutions.

    Prevention is better than cure, and it does not pay to pick the infection up!

If a large-scale trimming shows that heelhorn erosion control is desirable, the farmer would be well advised to use foot baths—and, if possible, to ensure a period of grazing.

Chapter 5

NUTRITION

In intensive dairy farming, nutrition has a considerable effect on the claws in that it is closely connected with diseases that in turn may be accompanied by laminitis.

DISEASE AND LAMINITIS

Clinical pictures (syndromes) that are caused or enhanced by nutrition play an important part in intensive dairy farming. Laminitis, varying from acute to chronic, may occur in a number of these syndromes.

In this connection the following disorders may be mentioned.

— Disorders of the **digestive tract**: rumen acidosis, inflammations of the digestive tract, poisoning through mouldy food.

— Disorders of the **uterus**: retained foetal membranes, inflammations.

— Disorders of the **udder**: udder oedema, inflammations.

— Acetonaemia.

Faulty nutrition is solely or partly the cause of these diseases, which in turn are held partly responsible for the origin of laminitis.

Laminitis may in principle occur at all times and in all ages of cattle, including young and store cattle. In store cattle rumen acidosis through overfeeding is the main cause of laminitis.

In *dairy cattle* it is noticeable that laminitis and related diseases occur particularly during the period of calving, from several days before calving until some three weeks after.

It is in this period that dairy cattle are often subject to a feeding regime which overloads the digestive tract and metabolism:

—Too many concentrates are often given before calving, and after calving the daily ration of concentrates is increased too rapidly. This increases the likelihood of rumen acidosis, udder oedema, udder inflammation and acetonaemia.

—The consumption of roughage is frequently too low at the time of calving. During this period total food intake is low and if a large amount of concentrates is supplied there is little room for roughage. This applies to heifers and over-weight animals in particular. Too little coarse-fibred roughage means that rumen contents will be poorly structured, and this is conducive to digestive disorders.

It is also increasingly the case that roughage is short of coarse fibres but rich in carbohydrates, as is the case with maize silage, young grass, short-cut hay, short-cut grass silage and roots. These feeds do not combine well with high intakes of concentrates; there is always the potential for digestive disorders.

— If a cow is too fat when calving, on account, for instance, of too high an intake of concentrates before calving, the chances of retention of foetal membranes, womb inflammation and acetonaemia are increased.

Our knowledge of the origin of laminitis is still very incomplete. The origin is chiefly sought in the presence of *poisonous compounds*, directly or indirectly affecting the metabolism in the corium.

The main source of poisons in the body is the digestive tract. In the case of digestive disorders the production of these poisonous substances, such as food derivatives and poisonous products of bacteria (e.g. lactic acid and histamine), could be increased.

Rumen acidosis is considered to be the main cause of laminitis. It will readily develop if the daily ration contains a lot of easily digestible carbohydrates (starch, sugars) and little roughage. Such a ration provokes little rumination and the food quickly passes through the rumen. Digestion is thus displaced to the intestines, which become too heavily loaded. This increases the chance of abnormal conversion in the bowels and the absorption of poisonous compounds from them.

A low level of rumination means low production of *saliva* which is necessary in the rumen to neutralise the acids (acetic acid, propionic acid and butyric acid) and prepare them to be absorbed. If these acids accumulate in the rumen, its contents become acidic and form *lactic acid*. Once absorbed, part of it is badly converted, so that the blood becomes too sour, and this could be conducive to laminitis.

A lot of carbohydrates, taken within a short period of time, increase the chance of acidosis of the rumen if insufficient roughage leads to a poor structure of its contents and inadequate rumination.

Sudden **changes in the daily ration** make acidosis of the rumen more likely.

Poisonous components of **mouldy fodder** might cause laminitis; several of these have an allergic effect.

PREVENTION OF LAMINITIS

As far as nutrition is concerned, the prevention of laminitis should be based on a **feeding regime adapted to the animal's need**. Only then will digestion be optimal and the production of poisonous substances be at a minimum. The cow stays healthy! Beware of **sudden changes in ration** and **mouldy fodder**.

These measures will be of benefit both to milk production and the duration of the cow's life.

They can be described as follows:

• Do not allow heavily pregnant cows to become fat. Feed dry cows until the day of calving in accordance with the norm of 5–10 kg of milk; the need of the calf before its birth is, at most, equal to the quantity of nutrients needed for the production of 5 kg of milk.

This means that until the day of calving dry cows must be given little or no concentrates apart from good roughage. The consumption of maize silage and good grass may even have to be limited in this period to prevent overfeeding.

• Draw up a safe schedule for the supply of concentrates around the time of calving:

Until the day of calving, 0–2 kg of concentrates per day (2 kg once a day). On the day of calving 2 kg *extra* of concentrates (2 kg twice a day).

From the second day after calving 0.5 kg of concentrates more per day, until the norm for the desired production has been reached. The amount of concentrates fed should not be brought to a maximum before two weeks have elapsed after calving, *or at least three weeks* in the case of heavy levels of concentrates or when the roughage contains few coarse fibres. It is possible that even four to six weeks is advisable.

• Avoid sudden changes in the daily ration, particularly during the calving period. It is particularly risky to change to food which has a greater content of protein and carbohydrates and a lower content of coarse fibres such as changing from hay to silage, or even from the winter ration to protein-rich short grass.

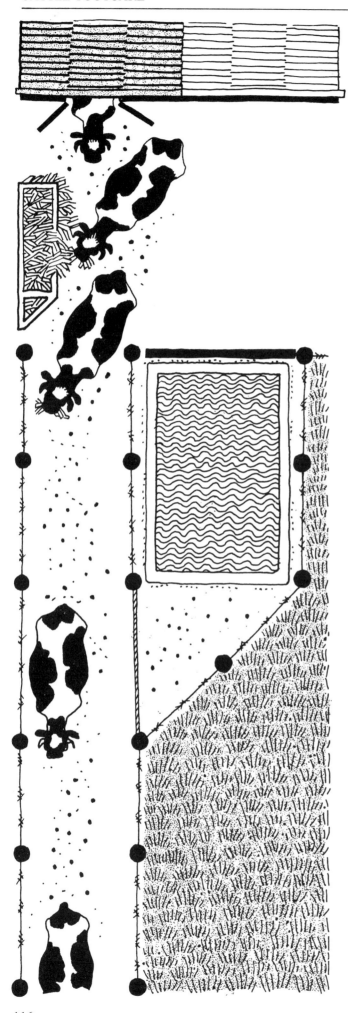

- Coarse hay or straw should be available around the time of calving if the roughage is deficient in coarse fibres; for instance if it is maize silage, short grass and clover, short-cut grass silage and short-cut hay, and roots.

  This advice applies to all dairy cattle during the grazing period under Dutch intensive farming conditions! . . . Through the foot bath to the collecting area, *past the hay-rack* back to the pasture . . .

- One-third of the total dry matter intake should originate from coarse-fibred roughage.

- Where food consumption is reduced (for whatever reason, but especially through disease), the intake of dry matter from roughage should be at least 60 per cent of the total consumption of dry matter. Concentrates may therefore constitute no more than 40 per cent of the ration.

- Feeds on which there is mould growth should not be given, particularly during the calving period.

  It is not accidental that claw problems increase as the feeding regime becomes more intensive. A reliable feeding system will help reduce lameness.

  If the herd is noticeably affected by laminitis, which might be the conclusion drawn as a result of a large-scale trimming, the question of diet should be considered carefully, perhaps in consultation with specialist advisers.

Chapter 6

HOUSING

What have housing conditions, including the organisation connected with it, to do with claws and claw lameness? Here the discussion is confined to the kind of dairy cow production where conditions are comparable with those in the Netherlands, involving various types of tied and cubicle housing.

The effect of housing may be all or nothing.

HOUSING CONDITIONS and:

— Disease

In one sense housing has nothing to do with claw lameness in so far as the lameness problem is essentially always the same from the veterinary point of view. This means that similar claw diseases and lameness occur in all sorts of housing. Although the incidence and seriousness differ from farm to farm, these differences appear to be connected with hereditary factors and farming practice – including organisation, feeding and care – rather than with housing conditions.

In our climate heelhorn erosion occurs everywhere. Laminitis is an inherent part of intensive and high-production dairy farming. The lameness problem has in fact little to do *with the type* of housing available.

Loose-housing with strawyards might have some advantages in this respect, but it is not experienced in the Netherlands.

There is, however, a connection with the *management of the farm*. For instance a system which involves keeping the cows indoors rather than at grass will favour heelhorn erosion. Heelhorn erosion will readily heal when the cows are out on grass. The degree of infection is reduced, which gives the claws a chance to recover to some extent from the malformations incurred in the previous housing period. This is clearly shown by the formation of new and healthy heel horn. If the cows are thoroughly trimmed at the beginning of the grazing season their claws will be restored to normal by the time they are housed at the end of the season. Whatever the housing system, reducing the time at grass means less chance of natural recovery and, certainly in the long run, a greater chance of problems . . . unless foot care is improved!

Reasonably dry housing conditions are less conducive to heelhorn erosion. To a certain extent, this has to do with the herdsman rather than with the type of housing.

Under extensive conditions where there is little housing heelhorn erosion will hardly ever be a problem.

Little is known about the relationship between laminitis and the type of housing. If the housing period is extended the prolonged influence of overloading will promote laminitis. Loose-housing with strawyards might be favourable in this respect.

—Lameness

The type of housing has nothing to do with lameness insofar as the systems we have at our disposal all have the same unnatural effect on weight distribution in the foot. Any hard surface will enhance the origin of typical sole lesions, laminitis and sole ulcers.

Housing conditions have everything to do with lameness in that the same lameness incidence is much more of a problem for the farmer operating a system of cubicle housing with central milking, than it is for one using tied stall accommodation and movable bail milking. Whether the farmer comes to the cow or the cow to the farmer may make a big difference; if not for the cow with good, healthy claws, certainly for the cow with bad, diseased claws.

The immediate effect that the farming *organisation* may have on the problem of claw lameness is related to the *demands* made on the moving mechanism of the cow, and the difficulties this entails for the *healing* of the ordinary claw-lame cow. Central milking exposes the lameness problem more clearly.

The tender cow, suffering from contusions, but not yet from sole ulcers, causes few problems in the *tie-up stall*. The tenderness may be noticed or it may not, her condition may deteriorate somewhat and she may be given some extra fodder, but daily work is not affected and milk production will for a long time go on as usual. However, trimming is advisable.

The lame cow, suffering from more serious contusions, an open lesion maybe, will be noticed. Thorough trimming, the sooner the better, will generally prodluce immediate results. Healing will go smoothly, as the claws have some rest and the lesions do not receive external injuries. The cow's condition and her milk production need not have suffered seriously at this point. In fact daily work is not impeded.

A seriously lame cow, however – a neglected case – always causes trouble: extra care will be needed if follow-up treatment is required, the cow's condition will decline and milk yield fall . . . perhaps the herdsman will be annoyed at his own negligence. Yet, even now, the farm's organisation experiences few difficulties.

The above applies whether the cows are housed or at grass, when the farmer goes to the cow for milking.

The situation will develop differently in a *cubicle-housing system*, as follows: A cow that is severely lame not only produces the problems described above, but also will no longer be able to walk to the feeding area and milking parlour easily. There will also be extra difficulties in the grazing season because longer distances will have to be covered and the tracks may be in a bad condition. The cow cannot keep up with the others and the process of healing may be seriously impeded if she is neither allowed rest nor able to avoid injuries from outside. The patient should in fact be separated from the herd, but organising separate feeding and milking will upset the daily routine.

If a cow is moderately lame, suffering from contusions and perhaps a simple open lesion, she will be noticed at once. The cow will stay in her cubicle and eat and drink too little. In some cases the cow will have to be led to the milking parlour. Additional care needed for one or more cows will hinder the running of a smooth organisation.

One advantage of an easily noticed lameness is that it will be given treatment at an early stage. Any open lesions may not get the rest they actually need, but moderate cases usually heal without too many difficulties. In light cases blocks will not always be necessary, at least while the animals are housed.

In the grazing season things are different: distances are longer and tracks are often in bad shape. For a healthy, intact claw this is no problem, but it causes difficulties for a diseased and defective claw. In this case blocks will have to be applied to ensure that healing is satisfactory. For one single cow this may not be a drawback, but treating several cows in this way will become quite a job. Furthermore these animals will also have trouble in keeping up with the rest of the herd; the same holds for the tender cow.

One and the same pathological condition will, in different farming systems, cause different problems and concerns. Cubicle housing and central milking expose the lameness problem more clearly. In fact lame cows can no longer be tolerated in 'modern' dairy farming: health care will have to be improved or selective breeding introduced.

An advantage of cubicle housing is that the cow with tender claws will be noticed earlier. This promotes timely large-scale trimming and its preventive effects.

At the same time rough yards and roads must be smoothed; small stones are disastrous for weak claws. In poor conditions external damage increases the development of a lameness problem and makes it more visible.

— Trimming

In trimming, housing conditions as well as farming organisation should be considered. Weather conditions also play a part.

After a lame cow has been treated an attempt must be made to adapt her living circumstances to the requirements of follow-up care.

On the other hand, *large-scale trimming should be adapted to the farming conditions*, as the farm must be kept functioning.

In claw trimming the *aim* is always the same: to attain weight division in and weight distribution over the claws as effectively as possible while *at the same time keeping their protective function intact*. Two, often conflicting, interests have to be safeguarded.

Sometimes one of the two functions outweighs the other. This **balancing of interests**, or defining of priorities, is a feature of claw trimming.

In the tie-up stall, the first priority is the loading of the corium of the outer hind claw; paring an outer claw sole somewhat thin is not too bad in this case. In cubicle housing more care must be taken in paring outer claws, although claws that are used to the system can be well trimmed. Under cubicle-housing conditions few walls normally have to be cut and most inner claws should be just scarcely trimmed.

For cows that are being cubicle housed *for the first time*, the protective function should certainly have priority. Trim judiciously – *do not trim at all in the toe region!* At first the rate of horn production has to adjust to the greater wear caused by the changed conditions; this 'critical period' may last 4–6 weeks or longer.

On soft grass, protection being less important, every attention can be paid to reducing the load on the outer hind claw; however this is just the situation where the problems of weight-bearing happen to be fewer. When the ground becomes hard during dry weather the protective function becomes more important, as does the unloading of a weak claw. One should try to find the golden mean.

The protective function is indispensable if the cows have to walk on yards and roads, to and from the milking parlour. An uneven hard surface, in particular, requires a firm sole to protect the quick. Leave the inner claw high and do not make the outer claw too thin!

The happy medium may be a useful rule for the chiropodist. However, obvious cases of lameness often require extreme remedies: open lesions need a clear, vigorous treatment (if necessary completed by a block).

N.B. Where cows are cubicle housed, unskilled trimming, e.g. cleaning up heelhorn erosion without ensuring a stable supporting surface, can seriously upset heat detection; tenderness influences normal behaviour.

If no lameness problem is present, large-scale trimming can produce it when a strong protective function for the claw is required by the farming conditions.

— Hygiene

Modern intensive dairy farming requires a certain degree of hygiene, not only in the milking parlour, but also in the living accommodation. Where a large number of animals are brought together in a relatively small area the conditions are right for the occurrence of bacterial and parasitic ailments. Regular cleaning and disinfecting (not only to fight heelhorn erosion) are required in order to prevent any level of infection in the animals' environment becoming too high.

In cubicle housing the chances of heelhorn erosion contamination are considerably increased, whereas in practice the annual cleaning of such housing is often neglected.

If the demands of hygiene are to be taken seriously, the consequences for the construction of the housing, including at least the *possibility* of achieving an annual or twice-yearly cleaning, must be considered.

Similarly construction and organisation must make it possible for foot baths to be applied simply and effectively. This should be planned before the construction starts and possibly in consultation with experts. (See also Chapter Four, for dimensions and location of the bath.)

Curiously, it sometimes happens that in the first year that cubicle housing is brought into use, large-scale bacterial problems occur in the feet, which in later years and under a similar hygienic regime 'automatically' decrease. So far the only explanation is 'difficulties in adaptation', a notion as vague as it is comprehensive.

Adding agricultural chalk (about 3 kg per cubicle) to the cubicle floor cover *might* have a favourable effect on the bacterial balance in the house.

—Fore Claws

Housing conditions as well as management affect the spreading of heelhorn erosion. In cubicle housing the fore claws are increasingly affected, particularly if the grazing season is continually reduced and bathing is neglected. This may result in the formation of fissures and in inefficient growth of horn, which make it desirable that the fore claws are trimmed. This seems an unwelcome extension of care.

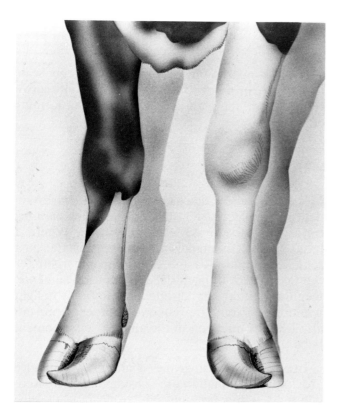

Furthermore, in cubicle housing it sometimes happens that a growing number of inner fore claws show excessive growth of horn. This is accompanied by a toe-out posture of the legs. In more distinct cases the animals have a somewhat jolting gait, but are not lame. This phenomenon could be related to the fact that, when they are feeding, the animals lean against the manger, their legs wide apart, having to stretch deep and far to reach the fodder. It follows that the inner claws are variably overloaded – comparable with the weight-bearing by the outer hind claws – which in turn would provoke excessive and aberrant horn formation. This problem can be overcome by raising the feeding gallery 20–30 cm.

In obvious cases trimming appears advisable. If this is needed on a large scale it means an extremely unwelcome extension of care in dairy husbandry. Observations of the behaviour of cows might perhaps reveal more insight into this aberration as well as suitable ways to prevent it. (N.B.—when they are grazing, cows place the fore legs in front of each other.)

ADJUSTMENT AND THE CRITICAL PERIOD

In new cubicle housing or on new rough road surfaces claws *may wear through* as the result of a combination of factors, which can be avoided. However, one of the factors cannot be prevented: it takes time for the rate of horn production to adjust to a change in conditions of wear. The quick needs 4–6 weeks (or longer) for this adjustment; after this time and under passable conditions, growth and wear will be balanced. Before this, wear will exceed production!

Conditions in cubicle housing may generally be considered passable for the dairy cow's claws if they are reasonably cared for; which means if they are reasonably healthy.

In the period of adjustment (the critical period) a claw may wear through due to a combination of the following factors:

- the claws are trimmed too short;
- the floor is too roughly finished;
- the cows have not been dehorned—they butt each other and make sudden movements;
- 'trudging about', due to unrest under unfamiliar conditions, is not checked.

Often it is in the toe that the sole wears through and in several claws at the same time. There are cases of extremely severe lameness and a number of animals may need to be culled.

If possible, trimming must be avoided during the three months prior to the use of a new concrete cubicle housing system or if it is likely that they will have to walk on new rough roads. In any case, the full length of the claw as well as the thickness of the sole in the toe area should be respected during this period! During the first two months in the new housing very little functional trimming can be carried out because of lack of horn.

In the Netherlands the problem of lameness in cows is not produced solely by accommodation and farming organisation. Housing conditions do have an influence but not decisively so. What is decisive is the care provided by the farmer or herdsman. If this becomes too heavy a task, the number of cows must be reduced or cows that are less susceptible to heelhorn erosion and laminitis must be bred.

120

Chapter 7

BREEDING

QUALITY OF THE CLAW – HEREDITY

It is often said that in order to reduce claw problems more attention should be paid in breeding to the quality of legs and claws. This supposes that this quality can be inherited.

It is necessary to distinguish between properties which are hereditary and those that are inborn but not hereditary. Hereditary properties are passed on by parents to offspring, whereas an inborn property with no hereditary influence is present from birth (at least in rudimentary form), but is not passed on to descendants.

In this chapter claws of 'good quality' is taken to mean those that rarely or never give rise to lameness and need little or no attention.

BREEDING AND CLAWS

Breeding is determined partly by the farmer and partly by the organisations that are at his service, such as A.I. organisations and herd-books.

The farmer may direct breeding on his farm by selecting the animals he will keep and by choosing certain bulls for his cows. The organisations influence this decision by their selection of sires and by their definition of the ideal cow or bull.

In their endeavour to breed cows with better legs and claws, farmers and organisations follow different routes. The farmer can act most efficiently by disposing of those cows which suffer most from lameness, or at least by disposing of their calves. This implies that he takes it for granted that at least part of the problem will be due to *hereditary* poor quality.

The organisations cannot apply this system, because they do not record lameness and even if they did they would have to take account of the differences between farms. Therefore they resort to an assessment of the quality of legs. This assessment is based on the posture and use of the legs, the structure and shape of legs and claws, and the presence of defects in claws or legs (such as concrete hock).

To make some progress in breeding several conditions must be fulfilled:

- The variety of the property 'claw quality' should be sufficiently large: there must be an obvious difference between animals whose claws are of good quality and those whose claws are bad.
- The difference in claw quality must to a large extent be based on hereditary tendency.
- The characteristics and data used for selection must represent the hereditary quality of the claws as far as possible.
- In order for a selection to be made there must be sufficient animals to choose from. The greater the number of properties that are included in the selection, the greater the number of animals needed to be chosen from.

No clear progress has been made in breeding as far as the lameness problem is concerned, partly because it does not sufficiently satisfy these conditions.

Dutch research into why cows are disposed of made it clear that 'only' 5 per cent of the animals removed were written off on account of lameness, the average age of these animals being higher than the average lifetime of livestock. This implies that there is also little *automatic* selection.

CHARACTERISTICS FOR CLAW QUALITY

In order for heritability of claw quality to be examined and used certain *characteristics* must be available that represent this property as well as possible; for instance the frequency of lameness, or the degree of lameness.

For that purpose it is important to determine the cause of lameness, because different causes may have different hereditary aspects (see below); certain cases of lameness, for instance, may be strongly influenced by accommodation and hygiene, and these are less suitable to be used as a characteristic for selection.

If characteristics for the selection like 'posture', 'structure' and 'shape' of legs and claws (sickle-hocked, cow-hocked, soundness of the leg, size of the claw, height of bulb, etc.) are used then it must be acceptable to assume that there is a connection between these characteristics and claw quality. The influence on these characteristics of other factors, such as housing conditions, must also be known.

No definite relationship between the evaluations of external features and the quality of legs and claws is known. Research is hampered by the difficulty of making these evaluations *uniform*.

In order to obtain identical and repeatable evaluations *measurable* characteristics have been sought:

- the dimensions and shape of the claw, such as width and sole surface;
- the chemical composition of the horn, for instance minerals or amino acids;
- the physical constitution of the horn, such as hardness and the ability to withstand wear;
- the microscopic construction of the horn.

Practical results have failed to materialise; it has not always been possible to measure the characteristics with sufficient reliability, and the relationship between the selected characteristics and the quality of the claw is uncertain.

HERITABILITY OF THE CAUSES OF LAMENESS

Little is known of the heritability of the common claw disorders. Most literature concerns interdigital overgrowth (tyloma), and some mention is made of the heritability of the 'corkscrew claw' and laminitis.

Research has been carried out on the heritability of **interdigital overgrowth**. Using the descendants of one pool of dams the incidence of interdigital overgrowth in the descendants of sires known for this problem has been compared with the descendants of other sires. Other investigations concern daughter groups of a number of A.I. bulls.

The results did not consistently reveal a clear hereditary influence. One possible explanation is that differences in findings could be caused by the occurrence of *different types* of interdigital overgrowth!

The prevailing opinion is, that there is a clear heritable factor in the case of the interdigital overgrowth situated exactly between the claws, covering the greater length of the interdigital skin; certain breeds seem to show this tendency. Overgrowths which cover only a part of the interdigital skin, often touching the outer claw (at least in hind legs) are thought to be caused mainly by external conditions (as there is continuous irritation and interdigital dermatitis).

Nevertheless certain animals or breeds are also said to be more susceptible than others to the latter form of interdigital overgrowth (or interdigital granuloma).

There are a number of abnormal claw shapes which suggest a hereditary influence. The most well known of them is the **corkscrew shape** in the outer hind claws.

Overall there is some confusion about the name and origin of these and similar abnormalities. It is generally assumed that the condition in the outer hind claws, if present on both sides, is inborn; heredity, however, has not been clearly demonstrated. Although on some farms descendants of particular bulls appear to show this twisted claw shape more than the descendants of other bulls, conclusions regarding heredity drawn from such data should be treated with some reservations.

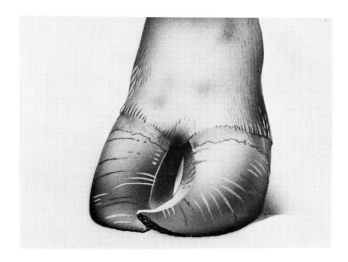

Lately it has been striking how similar abnormalities have developed in the *inner fore claws* of cows in cubicle housing systems. This condition may occur on a large scale and on both sides, whereas the young animals do not show any signs of abnormal fore claws. The scope of incidence varies widely from cowshed to cowshed. In the tie-up stall this phenomenon is unknown as a herd problem.

The cause may be multifactorial, but one cannot help thinking that housing conditions play an important role. (See also 'fore claws', pages 94 and 120.)

In accounts of **laminitis** an individual susceptibility for this condition is regularly mentioned. Those animals that have experienced an attack once are more readily affected again: however this may also be because the corium has been debilitated by previous attacks.

Sometimes it is said that 'certain families are more susceptible'. In a comparison between groups of daughters of A.I. bulls, differences have been observed in the occurrence of more and less serious forms of chronic laminitis. The hereditary influence did not appear strong enough for selection to offer many possibilities.

DIFFERENCES BETWEEN BREEDS

When cattle breeds are compared for lameness, the black-and-white breed appears very susceptible. In several investigations other well-known cattle breeds are reported to have done slightly better; but not a lot better. An exception is the Jersey breed, which seems to be considerably less susceptible to claw problems.

In the Netherlands, the red-and-white breed (M.R.IJ.) is said to have stronger legs and claws than the Friesian. In a comparative research on lameness in the red-and-white, the American-Canadian black-and-white and the Dutch black-and-white breed, the last proved to need veterinary care most, and the American-Canadian breed least of all.

CONCLUSION

Our knowledge of the heredity of claw quality and the causes of claw lameness is small.

We are completely ignorant of any hereditary susceptibility to interdigital phlegmon. In this ailment, accommodation and hygiene seem to be most important.

Some investigations point to a limited hereditary influence in interdigital dermatitis and laminitis.

In certain claw deformities, like corkscrew claw, and in some types of interdigital overgrowth hereditary influence seems to be of more importance.

In order for hereditary tendency to be demonstrated more needs to be known about the origin of abnormal conditions, to learn to recognise features that betray the predisposition of the claw. The role of environmental influences – such as housing conditions and feeding habits – on the onset of these features must also be known.

At present the only opportunity for the *farmer* to improve the claw quality of his own livestock is *selection on his own farm and based on the cases of lameness he has witnessed*. The requirements of his farm can be used as a guide: if management makes fewer demands on claw quality, selection is less important than it is in more exacting systems.

If good characteristics for claw quality are found in future, and if they prove to be sufficiently hereditary, then breeding organisations, too, may consider whether there is sufficient economic interest for selection to be based on them; in this case breeding for other hereditary tendencies such as milk production and shape of udder will have to wait.

EPILOGUE

THE MULTIFACTORIAL NATURE OF THE PROBLEM

One thing must have become clear through reading this book, namely that the lameness problem in dairy cows is a complicated and many-sided problem: a *multifactorial problem*. The answer is not one single solution, but a *multifactorial solution*. A start can be made with the most important factor or with the most easily affected factor; or, better, with several factors simultaneously.

Choices

Often the solution will have to include a *choice*, made by balancing the influence of the various factors; a choice between the pros and cons of the various possibilities; perhaps a choice between quantity and quality.

It is either more *or* better – more *and* better may no longer be possible. This is illustrated by the slogan of one of the great Dutch agricultural shows of some years ago: '*Not more but better*'.

The lameness problem is not a coincidental multifactorial problem, but a *structural* one related to the structure of intensive dairy farming; it appears as an '*occupational ailment*' of the cow, even prior to the use of cubicle housing.

Hess wrote in 1887 in his *Foot Diseases in Cattle* (translated from German): '. . . In the soft tissue of the claw . . . several diseases reside, which attack the larger ruminants, of different breeds, in the stable as well as on grass. . . . Increase of intensive dairy farming proportionately enhances these sufferings. . . .'

Haalstra writes under 'nutrition' that it is not a coincidence that claw problems increase if feeding is more intensive.

Real prevention must be found in improving the structure of dairy farming: nutrition, housing, breeding. This does not imply that effective measures within the present structure (trimming, bathing, hygiene), adopted in order to prevent worse symptoms, should be rejected as being palliatives. For better care means healthier claws and a longer lifetime.

We should, however, recognise that better care may result in a negative selection: the weaker animals will also have a chance!

On the other hand, in breeding, we must pay more attention to the claws: **a positive selection for claw quality seems indispensable for the future.**

Livestock science and veterinary science will have to co-operate.

I would like to draw the attention of experts on animal behaviour and housing conditions to the abnormal growth of horn in inner fore claws under the cubicle-housing system.

The Dutch farmer has been forced, or allowed himself to be forced, to increase his production. His labour budget reaches its ceiling and his financial budget is based on borrowed capital. What is the use of a farmer milking his cows for the bank? What is the use of a country passing her cattle breeding into the hands of the financial industry?

Authorities as well as farmers are decisive agents in a development that also creates the lameness problem in cattle.

PLANNING FOR CHANGE

The transition from tie-up stall to cubicle housing, and in general the transition from one farming system to another, requires that the cow and the farmer both make changes and satisfactory adjustments. The healthier the animals and the more sensible the farmer, the easier the change and the smoother the adjustment. *But adjustment always takes time.*

Claws that are diseased and in poor condition will suffer in the more exacting housing situations, and will make only slow adjustment.

A useful preparation for the transition to different conditions is to make the claws as healthy as possible. This may be achieved by means of correct nutrition, foot baths and hygiene, and judicious periodical trimming. **This assumes that the farmer has enough time to pay attention to these matters** – understaffing thwarts improvement.

- The control of a lameness problem begins with trimming all milk-producing cows. Thus individual cases are prevented from becoming worse, and the nature and degree of the problem can be properly outlined.

- If necessary the treated cows must receive follow-up care until complete recovery is accomplished. Because aftercare is time-consuming, and bearing in mind also the requirements of selection, the worst cases should possibly be culled.

- Knowledge of the scope and insight into the *nature* of the problem must lead to the *planning* of nutrition, bathing, trimming and farm organisation.

  It is the insights gained into the relationships between different factors that lead to significant progress.

FURTHER READING

AMERONGEN, J. J. VAN. 'Een onderzoek naar de bruikbaarheid van een rubber klauwblokje ter ontlasting van runder-klauwen.' ('An investigation into the use of rubber shoe blocks to avoid undue pressure in the feet of cattle.') *Tijdschr. Diergeneesk.*, *105*, 403 (1980).

ANDRIST, F. *Huf-, Horn- und Klauenpflege.* Graz, Leopold Stocker Verlag, 1952.

BRUHNKE, J. 'Vergleichende Studie der statisch beanspruchten Lederhautteile des Pferdehufes und der Rinderklaue.' *Arch. wiss. und prakt. Tierheilk.*, *57*, 324 (1928).

CHELI, R. and MORTELLARO, C. 'La dermatite digitale del bovino.' *Proc. VIII International Conference on Disease of Cattle*, p. 208, Milan, 1974.

Colour Atlas on Disorders of Cattle and Sheep Digits. Result of the International Symposia on Disorders of the Ruminant Digit. Edition du Point Vétérinaire, 25 Rue de Bourgelat, F – 94704 Maisons-Alfort.

CORNELISSE, J. L., PETERSE, D. J., and TOUSSAINT RAVEN, E. 'Een nieuwe aandoening aan de ondervoet van het rund; dermatitis digitalis?' ['A digital disorder in dairy cattle; dermatitis digitalis?'] *Tijdschr. Diergeneesk.*, *106*, 452, (1981). 'Formalinebaden ter bestrijding van de stinkpootinfectie bij rundvee.' ['Formalin foot baths in the prevention of interdigital dermatitis in cattle.'] *Tijdschr. Diergeneesk.*, *107*, 835 (1982).

DAVIES, R.C. 'Effects of regular formalin foot baths on the incidence of foot lameness in dairy cattle.' *Vet. Rec.*, *111*, 394 (1982).

EGERTON, J. R. and LAING, E. A. 'Comparison of *Bacteroides nodosus* infection in sheep and cattle.' *Proc. Fourth Intern. Symp. on Disorders of the Ruminant Digit.* Maisons-Alfort (Paris), 1982. (Ecole Nationale Veterinaire, J. Espinasse.)

EGERTON, J. R. and PARSONSON, J. M. 'Isolation of *Fusiformis nodosus* from cattle.' *Austr. Vet. J.*, *42*, 425 (1966).

FUCHS, G. 'Aspekte zum Begriff der Groszanlagentauglichkeit-Klauengesundheit.' *M. heft für Vet. Med.*, *31*, 930 (1976).

GREENOUGH, P. R., MacCALLUM, F. J., and WEAVER, A. D. *Lameness in Cattle.* Bristol, John Wright and Sons, 1981.

GÜNTHER, M. *Klauenkrankheiten.* Jena, Gustav Fischer Verlag, 1974.

HESS, E. 'Klauenkrankheiten', in: Bayer-Fröhner, *Handbuch der tierärztlichen Chirurgie und Geburtshilfe.* Wien-Leipzig, Braumüller, 1913.

NILSSON, S. A. 'Clinical, morphological and experimental studies of laminitis in cattle.' Thesis Stockholm, 1963. (*Act. Vet. Scand.*, *4 suppl. 1.*)

PETERSE, D. J. 'Een mogelijke invloed van de voeding op het optreden van zoolzweren bij het rund.' ['Nutrition as a possible factor in the pathogenesis of ulcers of the sole in cattle.'] *Tijdschr. Diergeneesk.*, *104*, 966 (1979). 'De beoordeling van de runderklauw op basis van het optreden van zoollaesies.' ['Assessing bovine claws by the occurrence of sole lesions.'] Thesis, Utrecht, 1980. 'Prevention of laminitis in Dutch dairy herds.' *Proc. Fourth Intern. Symp. on Disorders of the Ruminant Digit*, Maisons-Alfort (Paris), 1982. (Ecole Nationale Veterinaire, J. Espinasse.)

RANFT, W. H. 'Die feineren anatomischen Merkmale der Zehenknochen, speziell der Klauenbeine, des Rindes.' *Morphologisches Jahrbuch*, *78*, 377 (1936).

ROZTOCIL, V. 'Zur Aaetiologie des Rusterholz'schen Geschwüres der Ballenlederhaut des Rindes.' *Acta universitatis agriculture*, Brno, *XXX*, 393 (1961).

RUSTERHOLZ, A. 'Das spezifisch-traumatische Klauensohlengeschwür des Rindes.' *Schweiz. Arch. Tierheilk.*, *62*, 421 (1920).

SIMON, G. 'Untersuchungen über altersbedingte Veränderungen der Klauenbeine des Rindes.' Thesis Zürich, 1963.

SMEDEGAARD, H. H. 'Contusion of the sole in cattle.' *The Veterinarian*, *2*, 119 (1964). 'Foot rot or chronic foot rot in cattle,' *The Veterinarian*, *2*, 299 (1964).

TOUSSAINT RAVEN, E. 'Determination of weight-bearing by the bovine foot.' *Neth. J. Vet. Science*, vol. 5, numb. 2, 99 (1973). 'Lameness in cattle and foot care.' *Neth. J. Vet. Science*, vol. 5, numb. 2, 105 (1973).

TOUSSAINT RAVEN, E. and CORNELISSE, J. L. 'The specific contagious inflammation of the interdigital skin in cattle.' *Vet. Med. Review*, No. 2/3, 223 (1971).

WILKENS, H. 'Zur makroskopischen und mikroskopischen Morphologie der Rinderklaue mit einem Vergleich der Architektur von Klauen- und Hufröhrchen.' *Zbl. Vet. Med.*, *Reihe A. 11*, 163 (1964).

ZANTINGA, J. W. 'A comparative radiological and clinical study of the typical lesion of the sole (ulceration of the sole) in cattle.' *Neth. J. Vet. Science*, vol. 5, numb. 2, 88 (1973). (Summary of Thesis Utrecht, 1968.)

INDEX

abaxial 14
acetonaemia 60, 114
adaptation 22, 120
axial 14
– prominence 24

bandages 98, 100
basement membrane 37
biomechanics 21, 27
blocks 100, 102
breeding 121
bulb 10, 11, 14

claw quality 121, 122
comparison of claws 85
contusion 46, 47
corium 11, 22, 37, 56
corkscrew claw 19, 69, 123
coronet 10, 14
critical period 119, 120
curative trimming 95, 105

'dead' horn 22, 90
dermis, see corium
differences between claws 18, 19, 26, 31, 85
diffuse aseptic pododermatitis, see laminitis
digestive disorders 60, 114
digital cushion 11, 22
digital dermatitis 68, 72
disease control 52, 64, 107, 114
dressings, see bandages

epidermis 11, 22, 36
excessive horn formation 42, 54

fissures 41, 42, 95
fixing the leg 79
flat claws 93
foot bath 48, 52, 68, 107
footrot 4, 38, 66
fore claws 34, 51, 61, 78, 94, 120
formalin 109
foul-in-the-foot 66
fracture of the sole 63, 95
functional trimming 78, 105
functions 13
– bearing 14, 20
– protective 19

germinal layer 22, 36, 37, 56
grinding 106

growth rings 10, 14

haemorrhage 46, 53, 62
handling of tools 81
heel, see bulb
heelhorn erosion 35, 39, 66, 107, 117
height 17, 85
horn formation 37
– lesions 46, 54, 91, 95, 105
horny shoe 8, 10, 14
housing 117

inner claw 14
interdigital dermatitis, see heelhorn erosion
– granuloma 67
– overgrowth 50, 93, 122
– phlegmon, see foul-in-the-foot
– skin 10, 15
– space 10, 14

lameness 71, 92, 118
laminitis 35, 53, 114, 115, 118, 123
– acute 61
– chronic 53, 54
– subacute 62
– subclinical 62
length 17, 83
lime (chalk) 68, 119
loading of the quick 23

multifactorial disease 38, 52, 58, 124

nutrition 60, 114

outer claw 10, 14
overloading 45, 54, 60

papilliform overgrowth 51
parturition 56, 59
pedal bone 11, 20, 23, 29
periople 10, 11, 14
poisoning 60, 115
posture of legs 32, 47, 93
pressure 21, 23, 44, 45, 46
protrusion of corium 48, 98

quick 8, 20, 22

retained foetal membranes 59
rumen acidosis 115

sandcracks 71
slurry heel, see heelhorn erosion
sole 10, 11, 14, 91
– lesions = imperfect or interrupted horn formation due to pressure and/or laminitis
– thickness 19, 84, 88, 90
– ulcer = advanced stage of sole lesion 31, 47, 48, 54, 58, 62, 95, 98, 105
stability 16, 19, 85
stage I (heelhorn erosion) 39, 40
– (laminitis) 56
stage II (heelhorn erosion) 39, 43
– (laminitis) 54
– (foul-in-the-foot) 67
stones 69, 119
structure of the foot 10–11
subcutaneous connective tissue 11, 22
superficial purulent inflammation of the corium 70
supporting surface 16, 85, 88, 89, 96

tilted claw 49, 55
trimming 5, 26, 47, 48, 50, 52, 64, 70, 75, 119
– functional 78, 91, 105
– preventive 103
tyloma, see interdigital granuloma
'typical place' 31
typical sole-lesion 24, 25, 31, 45, 75

udder inflammation 59, 114
– oedema 57, 59, 114
ulceration of the sole, see sole ulcer

varying weight-bearing 28, 60
vicious circle 9, 30, 46
'vulnerable spot' 24, 25, 31

wall 10, 11, 14
– ulcer 49, 55, 95
weight-bearing border 10, 14
– surface, see supporting surface
white line 10, 11, 15
– disease, = white line lesion, = imperfect or interrupted horn formation due to laminitis 53, 55

zinc 68

DIRECTIONS FOR USE OF RUBBER CLAW-BLOCKS

If **no** sufficient difference in height between the two claws of one foot can be attained by functional trimming, the height of the **sound** claw must be **raised** by means of a block. To achieve this, a rubber block can be fixed to the claw with the aid of hoofnails. Attention should be paid to the following points:

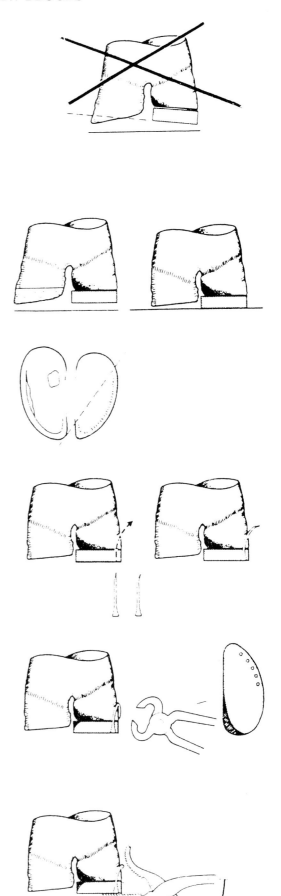

A. Trimming of the claw

The claw must be provided with a bearing surface which is **at right angles to the long axis of the shin bone**, but the sole must **not** be **thinned**. This can be realised by cutting the claw (flat and at right angles to the shin bone) in the toe part mainly on the outside, and leaving the bulb area as high as possible by limiting paring here to a minimum.

B. Fixing

Apart from the rubber block one needs:
— No. 3 hoofnails
— a hammer
— a pair of pincers
— suitable pliers

The nails must be inserted with the **bevel facing the sole.** The nail will then move outwards through the wall. The bevel side is indicated by a mark on the head of the nail.

The round nail-holes in the block allow the position of the nails to be fairly accurately determined. This is particularly important for the most posterior nails, as these should be directed slightly outwards. In the most posterior nail the bevel should be accentuated by bending the nail a little, in order to force the nail to leave the wall in the right place.

The nails must be driven into the so-called white line (just inside the bearing border of the wall) and leave the wall about 1.5 cm above its bearing surface. The ends of the nails are immediately bent downwards against the wall! A little grease on the nails makes it easier to drive them through the horn.

It is recommended to place the most posterior nail first, and then the most anterior one; or vice versa. 2 or 3 nails are then inserted between these two.

Attention: One should slope the upper edge of the block at the inside, in the bulb area. This is to avoid pressure by this edge of the block on the swollen bulb region of the 'diseased' claw.

The finishing touch is to cut the bent nails rather short (gather and dispose of the cut tips) and press the stumps of the bent and cut nails against the wall with the aid of the pliers.

Do not draw these stumps downwards through the wall. The bent stumps may be smoothed with a file.

C. Removal

The block should be removed after 3—5 weeks. The nails are driven somewhat deeper and the nail stumps are cut. The block can then be pulled off together with the nails.

Functional trimming is again indicated in order to load the affected and healing claw as favourably as possible.

The rubber block with nails is **not** suitable for short, worn claws (thin sole) and 'cork screw' claws.

FARMING PRESS BOOKS & VIDEOS

Below is a sample of the wide range of agricultural and veterinary books and videos published by Farming Press. For more information or for a free illustrated catalogue please contact:
Farming Press Books, Wharfedale Road, Ipswich IP1 4LG. Tel: (01473) 241122, Fax: (01473) 240501.

A Veterinary Book for Dairy Farmers
Roger Blowey

Deals with the full range of cattle and calf ailments, with the emphasis on preventive medicine.

The Principles of Dairy Farming
Kenneth Russell, revised by Ken Slater

The standard introduction to dairy farming covering the complete range of topics including buildings, farm systems, management, dairy farm crops and feed, milking techniques and milk production, breeding, calf rearing, disease control and profitability.

The Herdsman's Book
Malcolm Stansfield

The stockperson's guide to the dairy enterprise.

Cattle Ailments
Recognition and Treatment
Eddie Straiton (the 'TV Vet')

Recognition and treatment of common cattle ailments shown in over 300 vivid action photographs.

Calving the Cow and Care of the Calf
Eddie Straiton (the 'TV Vet')

Calving the cow and care of the calf covered in complete detail with 338 photographs.

Cattle Feeding
John Owen

A detailed account of the principles and practice of cattle feeding, including optimal diet formulation.

Indoor Beef Production
Ron Hardy and Sam Meadowcroft

Includes chapters on housing requirements, health, welfare, performance, cereal production, grass and maize silage, alternative feeds, finance, management and marketing.

Calf Rearing
Bill Thickett, Dan Mitchell, Bryan Hallows

Covers the housed rearing of calves to twelve weeks, reflecting modern experience in a wide variety of situations.

Forage Conservation and Feeding
5th Edition
W. F. Raymond, G. Shepperson and R. W. Waltham

Brings together the latest information on crop conservation, haymaking, silage making, mowing and field treatments, grass drying and forage feeding.

Making Profits with Dairy Cows and Quotas
Gordon Throup

Explains in full the principles of dairy farming as a business and the implications of quotas, followed by a detailed analysis of the various factors affecting profitability and efficiency.

Beef Breeds of Britain
Presented by Joe Henson, scripted by Val Porter

A VHS video which discusses the merits of traditional and newer British breeds, how several British breeds have developed in North America and some of the Continental breeds that have recently become familiar in Britain.

Cattle Behaviour
Clive Phillips

A complete account of the behaviour of cattle and its implications for welfare and production.

Footcare in Cattle
A VHS video in which Roger Blowey demonstrates hoof anatomy and links this to practical trimming techniques.

Stockmanship
P. R. English, G. Burgess, R. Segundo, J. Dunne

Gives a full account of the factors influencing the quality of stockmanship on the farm.

Improved Grassland Management
John Frame

Draws on the full range of current research to establish clear principles for the management of grassland.

Mastitis Control in Dairy Herds
Roger Blowey and Peter Edmondson

A clear account of what mastitis is and preventive measures that can be taken against it.

Farming Press Books & Videos is a division of Miller Freeman Professional Ltd which provides a wide range of media services in agriculture and allied businesses. Among the magazines published by the group **Arable Farming, Dairy Farmer, Farming News, Pig Farming** and **What's New in Farming.** For a specimen copy please contact the address above.